VISION

TO

VALUE

HOW TO STRUCTURE TECH COMPANIES

TO DELIVER GREAT PRODUCTS

LUÍS GOMES DE ABREU

www.visiontovalueframework.com

wydee

VISION TO VALUE

ISBN: 9781703015416 (paperback)
ASIN: B07Z956KY9 (ebook)

Published in December, 2019

Book design by Buro Blikgoed, Harlem, The Netherlands
Editing by Brandy Cross, Rotterdam, The Netherlands

wydee

Published by Wydee BV, Amsterdam, The Netherlands
www.wydee.com
www.visiontovalueframework.com

Introduction 11

Part I – A Mindset of Growth 19

The Journey From Vision to Value 21

Your Role in the Process 31

Part II – The Vision to Value Framework 39

Introducing the Framework 41

Vision and Strategic Plan 49

INTRODUCTION

So, you're a tech company?

Today's entrepreneurial landscape is booming with new ideas, concepts, and technologies. We have more opportunities than ever before to solve problems, create new things, and to do so in new and interesting ways. As a result, more than 3 million businesses are launched each year, each with their own ideas, visions, and goals. These might be solving a problem, creating a solution, or delivering a service. Amazon.com wanted to make books more accessible and affordable. Facebook wanted to connect people online.

Today, unlike any other time in history, vision is often realized and implemented in the form of a software product. Amazon's e-commerce store is built on software – as is Facebook's social media platform. And, most don't intend to realize software products when they create vision.

Despite no real goal or drive to become a software company and no real idea of how a software company works, many companies now create and deliver software products to consumers, either directly as a product or indirectly as a service.

Just 15 years ago, this evolution from service-company to software-company would have been unusual. Today, it's the norm. In 2011, Marc Andreessen penned his now-famous article for the Wall Street Journal "Software is Eating the World"[1] Today, the top 5 companies in the world are software companies. Amazon, Facebook, GE, and Exxon primarily develop software – although they use that software to deliver other services.

At the same time, many software companies don't operate like tech companies. Software requires a focus on long-term development and delivery to successfully manage and maintain software services over time, which physical products simply do not require. This necessitates a shift in focus from short-term product development to developing an infrastructure and process capable of supporting a tech company.

For example, operations, finance, and customer service are integrated into the product in SaaS companies. This intrinsically changes how organizations can and should operate.

In this changing environment, how must organizations design operations to enable delivering continuous value to the customer?

Why did I write this book?

Millions of new businesses launch each year. Most of these go on to fail within the first 1-5 years. With issues relating to market research, infrastructure, inability to scale, and lack of business processes, new businesses are simply, by design, error prone.

Startups are often fronted by people who have great ideas, but no experience

---

[1] https://www.wsj.com/articles/SB10001424053111903480809045765122250915629460

making tough business calls, no financial management experience, no branding, no business processes, and no real infrastructure. When these people go on to tackle large and complex business problems, they very often, and very naturally, fail.

While seed programs and accelerators exist to help these businesses get off the ground and reach a minimum viable product, many don't make it further.

Working as the COO of a scaling tech company, I often found there were no existing frameworks I could use to grow my organization. Instead, I was forced to "reinvent the wheel" multiple times, spending countless hours creating strategy to reach my goals.

My journey in the tech world began just after I graduated as a software engineer in 2006, when I started working for an internet company developing a CMS platform. Not long after, I met Michiel Chevalier, who had a vision – developing tooling to help Payroll and HR professionals streamline work processes and better serve their customers and employees. For me, this was a great opportunity to develop an ambitious platform from the ground up and our adventure began. We became co-founders of Nmbrs in 2008.

At the time, I had no idea what I was actually starting, or what challenges I would find along the way, which, I suppose, is what made it so compelling.

In 2009, we launched on the Dutch market with a well-defined product vision and market fit. At that time, our software processed 5,000 pay slips per month and our team consisted of just 5 people who mostly worked in payroll services.

As we scaled out of the startup stage, my role changed from that of software engineer into head of development. I switched focus from active development to guiding the company toward realizing its vision. I also took on responsibility for operations because, in tech companies, these concepts are intrinsically intertwined.

During our scale-up journey, we grew into an organization with 100+ employees, operating offices in Lisbon, Portugal and Amsterdam, The Netherlands, and processing over a million employees monthly for thousands of customers. Our industry entails complexities surrounding local regulation and labor laws, so our international expansion made it increasingly important for me to manage and build the infrastructure to support our growing company.

I researched a great deal and discussed my problems with peers as I worked to achieve this. There was no complete end-to-end solution describing the journey or the 'bumps on the road' which I and others were facing. My organization had to build everything from the ground up, even when other companies had already tackled similar problems.

Eventually we began researching solutions. We took an Agile approach of using small steps, testing everything, and using trial and error to determine what worked. Over time, and with research and experimentation, we began to find a consistent and replicable model. This model has been combined into a framework that orchestrates processes, teams, and people. In this book, I describe our framework in a way that can be used for development by organizations.

I learned a lot from my experiences with Nmbrs. However, success doesn't always highlight key factors of that success. Some of the most valuable information I learned came from situations that didn't work. It's much easier to understand why something doesn't work than why it does. In fact, I co-founded two startups that were not successful: Grappster, a product to help companies become more data-driven, and WallyLabs, a product that delivers social media, sales, and support dashboards to TV screens in commercial areas. While these companies failed, those experiences allowed me to pinpoint where I went wrong.

While writing, I am reminded of the fact that tiny details and processes, which are frequently overlooked, are often the key to success. In the bigger picture of product development, something as small as using the proper channels for communication, the right documentation platforms, or integrated systems, can be the figurative oil in the machinery that keeps everything running smoothly.

Taking steps to ensure you and your teams properly understand these details. Using processes such as team kickoffs to make sure the purpose or vision of a project is clearly understood and using retrospectives to find better solutions can determine and will have a great impact on the quality of your end-product. Your team handles the magic of product development. Your providing structure to facilitate that will allow them to focus on the most important thing – delivering value to the customer.

Here, my primary objective is to facilitate the journey of anyone starting or already in the process of scaling, so that such people can focus their efforts and creativity on developing big things for society and humanity, without the need to reinvent the wheel for their organization in the process. So, in a way, this book is for a younger version of myself.

While this book is intended to function as a framework to help your organization establish the grounding and guidelines to scale, it isn't about how your organization can grow from 10 to 200 people (or 200 to 1,000). It's also not intended to tell you what to do when designing and running your company. Instead, my goal is to provide a foundation for structuring technical operations and to create solid and sustainable strategies for developing and continuously deliver a successful software product. Rather than giving you a recipe for success, my goal is to create a framework which anyone can use to make the right choices for their organization.

It's important to understand that strategies are not working solutions. They can guide your company and provide inspiration, but no strategy or operational framework will intrinsically fit your company. It's crucial to continue to revise and update strategies based on external and internal changes.

Most people understand that 100 degrees is the boiling point for water. However, if you go above or below sea level, this temperature changes based on atmospheric pressure. At an altitude of 2,000 meters above sea level, water boils at 93.4 degrees Celsius.

The conditions for change vary dramatically depending on factors such as the environment, audience, and even your organizational goals. There is no one recipe for success that brings everyone to the top. It's crucial that you make your own decisions based on organizational capabilities, goals, products, budget, and the market. You can and should adapt my framework and strategies to your organizational goals and needs to ensure it is a good fit for your specific situation. My recommendation is to approach any strategy or framework like a tool. It can help you to get where you want to be, but you must review whether it meets your needs and make changes to ensure that it continues to meet your organization's needs.

I also wrote this book with the intention of creating a broader and more comprehensive framework for technical companies. While there are many frameworks aimed at building technical operations or creating agile business structure, many fail to focus on the broader scope. Most agile frameworks such as Less, and Disciplined Agile are oriented at a team level, describing practices and tools such as retrospectives, standups, and pair programming. This focus is often too technically oriented or too focused on specific output. Therefore, it is not useful for developing an organizational structure capable of supporting software delivery – where teams, budgets, stakeholders, and other relevant parties must be taken into account.

While Agile is important – and I incorporate many of the key concepts of Agile in my own framework – I believe it's important to recognize the bigger picture. For this reason, I've shifted my focus away from purely technical operations to a much broader scope focused on structuring vision with a holistic technical view of operations.

In summary, I wrote this book in hopes it will help you when designing structure for a better and more sustainable tech company.

Using strategy for decision-making

This book covers many aspects of tech product development, ranging from operational processes for software delivery to team organization and value-stream mapping. This information consists of strategy, which you should use to guide your decision-making. Therefore, I have placed considerable emphasis on using strategy, not just when setting up your organization and making key decisions, but also during day-to-day operations, which are just-as-if-not-more crucial, than major decisions.

Organizing, setting up, and running a tech operation requires that you make numerous decisions, taking dozens to hundreds of tiny details into account. You will be faced with situations where you have to make decisions quickly and correctly at a moment's notice. Whether you delegate or work in a team, people will ask for support and guidance. Understanding the importance of using strategy to make those decisions will help you to guide your organization as you scale.

While you can and will often have to make decisions on the spot, these decisions can very easily be wrong if you don't use the proper guidance. Depending on the moment, your mood, stress, or any circumstance, you may make a wrong decision or make a decision that is inconsistent with your goals. The best answer is to rely on strategies, so you know every decision is consistent and aligned with your goals.

When someone asks, "What do we do in this situation", you can ask "What's the strategy?".

Once you define your strategy, you can use it to make consistent decisions based on goals and ideas. This will help you when planning, but will also help your team in making decisions, and in supporting you based on whether decisions and actions are aligned with the strategy.

Strategy is an exceedingly important part of any operation, but especially so during periods of growth and scaling. With this in mind, I have suggested several strategies throughout this book. You may not agree with every strategy I propose, but that's good! My strategies may not work for your organization. Instead, I recommend that you develop your own strategies or adapt mine to your needs if you use them to guide your work and decisions.

How is the book structured?

Writing this book, I was faced with the immensely difficult task of formatting information in a way that would offer the most value. I eventually chose to break the book into three parts, beginning with Mindset, moving into the Framework, and then following up with more information on practical application. In the following paragraphs, I will briefly discuss those points to give you some idea of what lies ahead.

Part 1: A Mindset of Growth

In A Mindset of Growth, I introduce my Vision to Value Framework, its scope, and its owner inside of your organization. In this chapter, I've worked to present the theoretical and practical knowledge which forms the foundation for the rest of the book. This includes the "journey from vision to the value", your or the tech COOs role in the journey, and information on why I believe a mindset of growth is crucial for modern tech companies. This section will not tell you what the framework is about, but rather, why it is important, what it influences, and how it impacts your organization as a whole.

Part 2: The Vision to Value Framework

In "The Vision to Value Framework", I discuss the framework in detail, how it relates to your organization, and its scope. This section introduces the framework, its parts, and how they fit together. This will give you the basis to begin applying my framework for your own organization. This section also includes information intended to guide you or to serve as inspiration for strategy, which you can choose to apply inside your own organization. My approach is to discuss the pillars of the framework, how they relate to your organization and its business goals, and offer tips and case studies from my own organization for implementation.

Part 3: Structuring Tech Operations

Structuring tech operations offers insight into the practical elements of setting up product development structure and developing business processes. Here, I share my experience with designing structure, organizing teams, and creating an operational framework designed to deliver value. My goal for this chapter is to help you realize a framework inside your own organization.

PART I

A MINDSET OF GROWTH

Scaling any organization is a challenging process which requires massive internal change. As your organization grows out of the startup phase, its focus shifts from simply surviving to building processes that will support a sustainable organization. As the Tech COO, your role is often to steer the company towards delivering on the CEO's vision, while creating operational strategy to enable long-term success. In this chapter, we will discuss the foundational knowledge and mindset required for the rest of the book, so that you understand not just what, but also why, your role impacts the organization.

The Journey From Vision to Value

"The journey of a thousand miles begins with one step"

- Lao Tzu

Nearly any tech organization will follow a defined path from startup to a sustainable and profitable organization. While not every startup will eventually become a sustainable and profitable organization, and many will take years or even decades to get there, the process is very often the same. I typically define organizational growth in three stages, each of which has its own period of growth and goals.

At Stage 1, the startup has a vision and an idea and works to develop a product to deliver that vision to the customer. This stage typically includes 1-20 employees and is a very technical stage, where organizations must focus on product development over anything else. However, the organization needs a minimum viable structure to achieve that product development.

The start-up phase of a tech company is very much focused on refining a vision. This begins with finding a business model and building a product MVP (Minimum Viable Product). In this stage, most companies focus on customer discovery, prototyping, validation, and refining products and features until the product reaches early-adopters. These users are very keen to try and use unfinished products and provide developers with valuable feedback for improvement and optimization.

This stage of growth is often a turbulent one. Tech startups require a significant amount of funding and many go into survival mode – simply working to put out a product while they can – rather than developing sustainable organization and structure. While understandable considering often-tough product development timelines and budgets, failing to develop minimum viable structure alongside your MVP will affect growth at a later point – and sometimes even your ability to deliver your first product. As a result, it's a very difficult stage for startups, who are often in a race against the clock to develop structure and a product with a limited budget.

Organizations making it through this first phase may be led to believe that everything afterwards is easy... but is it?

The second stage of organizational development, where companies begin the transition from startup to scale-up is also very challenging. This process typically starts when an organization employs around 20 people. It begins with the need to structure continuous technical operations that can push value to users in the form of software and services. Here, the organization has a defined vision, the product is running, and there are some early adopters or users. The company is still far from achieving its end goal. In most cases, this end-goal is to grow or 'conquer the world' – in whatever way that translates into the organization's playing field – by delivering value to the customer and increasing market demand. This is only possible with a great vision. Therefore, this stage is fundamentally different from the previous one.

You can picture the second stage as something of working with a blank canvas. You have all the tools you need, a selection of oil paints and pencils, and an idea of where you want to be. You could immediately begin to paint, adding layers to your organization. Or, you could take the time to sketch your structure, creating guidelines, standards, and a picture of where your organization should be, which will guide your operation as you grow.

Here, companies must begin the process of bridging the gap between the initial company vision and delivering value to the customer. While startups focus on developing a solution to deliver their vision with limited timelines and resources – producing something so long as its anything – scale-ups must switch focus to structuring and developing the organization for the longer term.

In the second stage, delivering value using sustainable operations is significantly more important than producing a product. Your organization must switch focus to consider not just what it is producing, but also how it delivers value to the customer and how it contributes to customer satisfaction. If your organization continues the race to produce a product and simply put out more features or better iterations of the software, it is possible it will miss that goal.

Achieving the fundamental shift to deliver value requires a process of scaling up to meet increasing demand, designing structure to ensure consistent and repeatable results, and setting up operations to enable the continued success of the company.

During the scale-up stage, most companies expand to 20-200 employees and must work to build sustainable processes that enable continuously delivering value through mixed technical and organizational growth.

Once your organization has this structure in place, it moves on to stage 3. Stage 3 is about ongoing growth, which entails optimizing structure and processes to continue to deliver value. This stage is very organizational rather than technical.

In stage 3, most organizations have a lot of processes in place. If you're reading this book for a stage 3 organization, your focus should be on tweaking and optimizing what you have rather than creating a new framework. Your organization's framework already exists, because organizations only reach stage 3 when everything is already in place. You cannot scale without that structure.

Throughout this book, I've focused on stage 2 rather than stage 1 and 3. Why? Stage one is about developing your MVP and minimum viable structure. In stage 2, you begin to focus on designing stronger operations that support further growth, which is where my framework offers the most value. In stage 3, you should normally have processes in place, and can use the framework as a validation and improvement tool. However, you can still use the Vision to Value Framework to optimize and streamline your operations or your future efforts.

The Vision to Value Framework shares several strategies useful in the second stage of organizational growth. With strategy for structuring technical operations for delivering software products in Agile environments, the framework helps ensure controlled and sustainable growth. I also touch on product development, customer support, startup/onboarding, sales, and marketing, because these greatly impact your operations.

The operational framework

Moving from stage one of startup and product development to stage 2 of growth is often difficult and many tech COOs begin to face challenges. Most startups allocate COO responsibilities to either partners or founders, who often don't have the necessary knowledge or experience to build structure for long-term organizational growth. This necessitates a substantial shift in approach, as individuals who start in technical or product-development roles must change their mindset to creating company structure.

Transitioning from realizing vision to creating a structure that consistently produces value is a process requiring time and experimentation to understand what truly works for your organization.

The first logical steps for any organization are always defining concepts such as "What is value" and working to determine market interest and demand. Once you've achieved these steps, you need to create something to deliver that value. And, once you have a product to deliver value, you need operational structure. For example, if your organization launches with the intent of developing an app to solve a problem, solving the problem is your vision. The app as a solution is the value. The first step to achieving the app is to create the code and processes that make up that app. Once you have this "product" in place to realize vision, you must create organizational structure to support the continuous delivery of that product or value.

Operations provide the basis for enabling larger companies to continue delivering value by creating structure, priorities, and strategy determining the future of the company. As tech COO, making this shift from focusing on realizing a vision during startup to enabling continued value during scale-up is a large part of your role.

Bridging that gap requires having a design or structure to support your business strategy and deliver your product as you scale. This design is known as an operational model. This book defines one such model, the Vision to Value Framework. This framework establishes the foundation to achieve core organizational goals and values such as enduring for the long-term, becoming customer-centric, and achieving organic growth. It will help you structure technical operations using approaches like focusing on end-results, sustainable growth, and incremental development.

Starting from value

Delivering value to the customer is the end-goal of any business. While the definition of value is very broad – I will touch on the topic later in this book – you can, in our context, simply define value as the deliverables you bring to the customer. This may include product development and services, quality customer service, and other factors such as the specific function of your product(s). Understanding value for your organization is crucial to creating an efficient operation because it enables you to understand how your organization delivers for the customer. End-value or final output is the key to doing so.

It's easy to get caught up in creating the perfect products, processes, and tools. Everyone wants to be the best at their job, and that means focusing on technical excellence. While this is desirable and can facilitate delivering a better and cleaner end-product, technical perfection or a total technical operations solution should never be your immediate goal. Your organization's goal is to build a product that delivers value to the customer.

To achieve this, you should establish a process that begins and ends with value. If you can define what value is to your customer, you can work backwards from it to determine how and what you need to do to achieve that value. This helps you switch focus away from simply building "something" and allows you to create a system entirely focused on producing your end-goal. It's about defining your ideal product (the end-goal and how you deliver value) and creating a structure that supports your organization's ability to deliver that product

This process is naturally circular because starting from value will influence what you build. This also means your solution will adapt as you build, which will further influence how you deliver value.

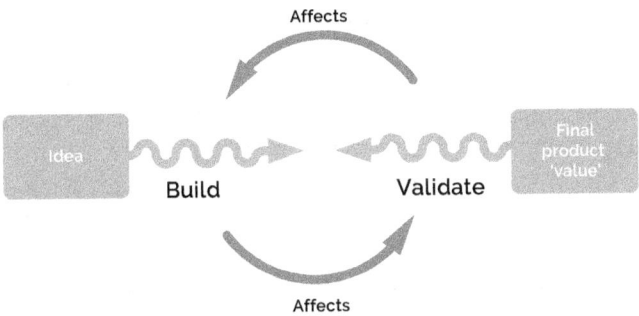

Fig.1. Working backwards from value

How does that apply in a practical environment? At Nmbrs, we were working to deliver new HR features onto the platform with a tight deadline. We distributed work across several teams to make that deadline, as customers were aware of the product's development and had been waiting on it for several months. The teams were therefore under a great deal of deadline stress.

As a result, each team was completely focused on their own pieces, and everyone was quite positive about their progress. Each piece was almost finished, so it seemed to be on track.

Then, we stopped to ask if anyone had put the different pieces together to perform a final use case. Silence. No one had stopped to check user experience. This was a big red flag for me. We quickly switched development focus to connect all the pieces in a pre-production environment, with frequent integration testing to ensure that the result would be a good one. Using this new process, we quickly discovered numerous tiny details which had been overlooked before and were now easily fixed on time for launch. Once the team shifted their mindset to focus on value for the end-user rather than technical details, we were able to align what we wanted to deliver and developing accordingly.

This is something that holds true for any technical company; it's crucial to keep the end-value in sight rather than getting caught up in technical details.

Sustainable growth

Launching a business is something anyone can do. In fact, more than 3 businesses are launched every second (more than 100 million per year). Developing a business into something that continues to grow is another question entirely. More than 70% of those businesses go on to fail. But what goes wrong?

In my opinion, most organizations make mistakes regarding planning for growth, for stability, and for the long rather than the short term. Your organization is growing, it is only natural that you experience growing pains. They are normal and part of the game. While the natural response to rapid growth is to create quick and temporary solutions, it is crucial that you take the time to develop operations for what I call sustainable growth.

One focus of this book is to help you do so, using strategies and approaches designed to create long-term benefits, so you can create a future-proof operational structure.

For example, businesses often fail when they don't have the structure in place to support scaling. Without strategy and focus, they can't maintain product features or access for a growing number of users. Without quality leadership, organizational focus, and customer service, organizations can't continue to succeed long-term. And, without structures in place to manage resources such as quality, funds, employees, and features, the organization will eventually fail to meet expectations. Many organizations start to scale before having these features in place, often because they don't realize they need them. This is a huge risk for startups receiving investment rounds, because seed funds mean they can grow more quickly than their organizational structure will support.

In Simon Sinek's "The Infinite Game"[2], the author introduces the concept of playing an infinite rather than a single-phase game. This concept applies extremely well to software companies developing SaaS products, because you must continue to develop and build for long-term success.

If a football game is 90 minutes, you play to have a goal advantage after 90 minutes, even if that means simply defending after having scored advantage. If the game is longer, you have to change your tactics and play to keep scoring because you must maintain that advantage for longer. Developing SaaS products is a lot like playing that game. If you plan your strategy for the short-term, score once with good product development and then defend, you won't be able to sustain continued growth and eventually you will lose to a competitor, especially as new players enter the field. Technical companies must shift focus away from short-term product development and towards a continuous cycle of development and feedback implementation, supporting continuous improvement and value for the customer.

This also ties into the concept of building for the long-term. One of the most interesting views of creating solid organizational fundamentals and a long-term vision is explained in the work of James C. Collins and Jerry I. Porras, "Built to Last"[3]. Here, the authors explain that most major companies like Boeing, Walt Disney, and IBM – who have thrived for decades – have a set of common principles which distinguish them from companies that do not last. These companies work to preserve their core vision in their DNA, with a strong vision and purpose and a culture developed around helping the company adapt and innovate over time.

The key defining factor of greatness is how these companies' approach what they do. It's their switch from "telling time" to "clock-building" that sets them apart. This means that truly successful companies aren't just good at specific

---

2   https://www.penguinrandomhouse.com/books/547570/the-infinite-game-by-simon-sinek/9780735213500/

3   https://www.amazon.com/Built-Last-Successful-Visionary-Essentials/dp/0060516402

things like manufacturing airplanes or making animations. In order to continue delivering value, they shift focus to building an organization that could grow and deliver customer value on its own, while remaining aligned with the founding principles.

Each of these companies achieved this shift to "clock building" by sharing the founder's mindset and purpose with the entire organization, so that everyone in the organization could become an extension of core organizational beliefs. This is achieved through select recruitment, intense onboarding, and the implementation of processes and practices following the company's core purpose and beliefs. For example, HP made this shift by developing the HP Way[4], which focuses company culture on people, innovation, customer centricity, integrity, and teamwork, developed into five basic tenets:

1. We have trust and respect for individuals
2. We focus on a high level of achievement and contribution
3. We conduct our business with uncompromising integrity
4. We achieve our common objectives through teamwork
5. We encourage flexibility and innovation

These principles were used to drive HP at its core, forming the heart of business decisions, employee onboarding, and leadership decisions, with the core message that employee's individuality and contributions were the company's most important resource. HP integrated tuition assistance, flex time, and job-sharing as part of the HP Way, during which time the organization doubled in size.

As a tech COO guiding your company along the journey from vision to value, you must function as a "clock builder" so that your teams will grow to deliver value themselves. This means recognizing what provides value to the customer, working out how to achieve that value, and creating organizational structure to support it.

This will become more essential as you continue to grow because creating a solid foundation provides the means to smoothly scale-up, without having to reinvent new solutions and processes each time you outgrow a previous rapid solution. My approach is to create processes in a way that generates positive improvement cycles, so that they improve and run more and more smoothly after start.

GreenForMe[5] is a German startup offering an online gardening service, helping customers to choose plants, pots, and a garden layout based on answers to simple questions like soil type, light, and climate. The company was founded by two people, one with a gardening background and the other with a business and marketing background, who applied for a startup incubator.

During the incubator program, GreenForMe received help with creating a business plan and launching their software product. They assembled develop-

---

4   http://www.inflexion-point.com/Blog/bid/74097/5-Timeless-Principles-Revisiting-the-HP-Way
5   https://www.greenforme.de

ers, UX and UI designers, and worked to launch their MVP (Minimum Viable Product). Despite having all the pieces in place, the product was never finished.

I was brought in for a session to advise the founders, assess the situation, and advise on technical operations. One thing quickly became obvious. The founders didn't realize they'd just started a software company. Their goal to develop, launch, and monetize their product simply did not work with software.

There's no such thing as a single product development cycle (develop, sell, make a profit) for software. Instead, product development and delivery go hand-in-hand, with business activities, where development is an ongoing part of the process. Once you are able to deliver an MVP, you receive feedback and begin to add new services and features, while continuing maintenance. Product development is the end-goal for product companies. For software companies, it's the beginning of the journey.

GreenForMe was missing that necessary shift, the focus on how they were delivering their product, not just the product itself. They were missing a good bridge between vision and delivering value.

In this way, many startup incubators only focus on the short-term success of the company. The startup incubator GreenForMe followed did not instill a strong sense of delivering value, which can be determinant in the success of early-stage startups, where every penny spent counts, and you can only spend it once.

While startup incubators often contribute to the early success companies developing software products, they do little to contribute to developing infrastructure for long-term sustainability. Few accelerator companies can build teams, find the customer base, or raise the capital necessary to scale for long-term success. With assistance and funding sometimes limited to just a few months, startups must focus on developing an MVP, so even minimum viable infrastructure often and unfortunately becomes a secondary priority.

Developing for continuous and sustainable growth is crucial for SaaS. This means making a conscious shift towards continuous development, changing your strategy for the long-term, and working on offensive tactics that will help you keep up, even as new competitors enter the field. As a startup, you should work to create the minimum necessary infrastructure to launch your product. As you move into stage two, your focus should be on infrastructure.

In this new environment, user satisfaction must be the end goal and your processes and structure. This ties into identifying your value and working backwards towards a point of value.

Setting the correct targets, responsibilities, and processes will help teams to increase their efficiency, take ownership of what they are doing, work autonomously, and work to deliver value to the customer, so that your business prospers.

Strategy Tip: Aim for long-term solutions rather than short term victories.

Evolution as a mindset

Any scaling organization will undergo complex and rapid changes. The process of scaling from 20 to 200 people is a large one and it will change how your company works from the inside out. Co-founders and CEOs will be forced into new roles, developers will move into roles as team leads, and the Tech COO position will evolve as you bring on specialist managers to handle many of your personal existing roles as the job becomes too big for one person.

That growth will also naturally challenge and change what your organization needs from an operational model. Why? Even with a fixed long-term company vision, the things your customers need and how your company delivers to them will evolve as your user-base grows, as demands change, and as technology changes. Your operational model will have to adapt to meet those needs, so that you have the structure and support for what is needed without investing too much.

I like to use a City Metaphor to describe this phenomenon, where I compare the growth and strategy of a tech company to that of a city.

A city starts as a small settlement, established for a reason or interest, and it begins to grow. As social and economic forces drive growth, the influx of people creates changes and the city is planned around that population. While it would be cheaper to design for the far future, it wouldn't be feasible or practical in the now. Designing a city like New York for 50,000 inhabitants would fail to deliver quality of life to the residents. Instead, streets have to be smaller, more accessible, and closer together. As the city grows, streets, buildings, and even neighborhoods must be torn down and rebuilt to suit changing needs. New resources like schools, hospitals, and public transit must be built as needed, not before or after.

A city is a complex system that requires planning and infrastructure designed for the needs of the population now. Planning for the future is important, but anyone could tell that building a 200-stop public transport system in a small town is a waste of time and resources. At the same time, the city is never finished, and never perfect. It must keep evolving with the needs of its population.

You can hopefully see how the "city" greatly parallels the growing tech company. Developing software and applications often results in investing in solutions no one needs, over-engineering, and otherwise developing solutions that are too big for what the customer needs now. By the time those solutions are needed, they could be too old, the wrong solution, or completely unfit for the customers now.

Instead, we have to go into a mindset that we need to keep rebuilding and reiterating the current solution. Just like a city must tear down a neighborhood or a transit network to improve, your company will have to focus on rebuilding and improving to keep meeting the growing and changing needs. This affects software and features as well as the internal structure and operational model. New versions and redesign are not a result of bad design, but rather smart operations.

Organic growth is driven by customer demand, which means that your focus must always be on practical and user-oriented solutions that meet customer needs now and in the near future.

A great deal of my work at Nmbrs has involved designing and creating a structure for product development, to ensure we could scale the product and the company. When we launched Nmbrs, we did not put all processes in place right away. Why? Part of the reason was that I came from a non-COO background and wasn't aware of the technical processes that should be implemented in a larger organization. However, my not doing so right away benefited the organization in the long-term.

Implementing processes and structures for a larger organization would have over-complicated the situation rather than enabling growth. Instead, we implemented processes as they became necessary, which kept work as simple as possible during our initial scale-up. As your teams grow, you have to coordinate roles, create processes to facilitate communication, define clear responsibilities, and invest in internal documentation. Implementing these processes at the start would offer very little in the way of value (except for internal documentation) but would get in the way.

If I were to scale an organization again, one of my biggest challenges would be to create an evolutionary timeline, defining which processes are relevant and should be introduced at which stage. Of course, understanding how larger organizations work will help you to define your own path for growth. And, that's one of the purposes of this book – to introduce you to an operational model that will help you design your own organization.

Strategy Tip: Don't focus on the final solution, evolve your solutions and implementations.

Your Role in the Process

"I'm not playing a role. I'm being myself, whatever the hell that is"
— Bea Arthur

Organizations moving out of the startup phase are faced with changing priorities, because growth necessitates a switch to an operational rather than a purely technical focus. As the COO under any name or title, your role spans a substantial portion of the functional areas of your growing company. You are responsible for giving direction and organization to the CEO's vision and you are often responsible for the technical operations and strategy necessary for realizing that vision. Understanding your role in the process and how you contribute to both the growth and the long-term sustainability of the organization, including how your role overlaps with and ties into others inside the company, will give you a clear view of how you fit and contribute to your organization's growth.

Most startups don't have COOs. The individuals who take on the responsibilities often do so with no real defined role, title, or even understanding of the necessity of the position. The role naturally varies from company to company, with different organizations requiring different levels of technical and operational organization. In some organizations, the COO takes on the role of anything that isn't yet filled, in others, it's a clear-cut and defined responsibility. As your scale-up moves out of the startup phase and begins to grow, the COO will naturally hire on or delegate new employees to take on facets of their original broad role. This is important as the COO role evolves to take on more responsibilities in specific areas, such as strategy, rather than being hands on with all areas of technical operation.

Despite the ambiguous nature of your role as technical leader, it is you who are responsible for guiding your organization into growth and hyper-growth by providing the operational strategy and organization needed to succeed. Your role is focused on enabling growth by helping the organization to create operating systems designed to share information, align priorities, and make decisions aligned with the CEOs vision and the company strategy. Communicating how and why the company is run, having the capacity to adapt your operational strategy over time, and creating an information and data system to ensure every key person knows what's going on and why is crucial to this role. While all of this is important, I will focus on a few key aspects of this process, because this book is about creating operational structure and not specifically your role as COO.

The technical COO role

Tech organizations in the scale-up stage already have a defined product and market fit. They know who they are and who they are selling to. They only have to accelerate the pace of product development while setting up or reimplementing services like customer support to meet growing demand in order to scale.

In tech organizations, the entire operation revolves around a software product. This means that the COO (Chief Operating Officer), who is responsible for operations, works in a role which is intrinsically linked to that of the CTO. In fact, many organizations combine these roles until the company reaches a certain number of employees.

While the role of CTO is often prioritized in startups, it makes a great deal of sense to combine the roles in tech companies, especially in the early stages of growth where there simply isn't room or need for both. Product delivery in tech companies is very much intertwined with organizational structure, because technical operations are essential and part of the process of software development.

However, there are still differences between the roles. As a COO, you are involved in developing company structure and collaborating with HR and other organizational teams. As a CTO, you would normally only be involved with technical vision, linking the product to vision, and production. As a tech COO, you must combine both roles, working inside both "layers". Why? In tech companies the two are completely intertwined and as a tech COO, you must work with different teams to build the technical capability and product scope of the company as well as the organizational structure supporting it.

Here, the core business is software product development, normally developed under the umbrella of CTO. However, tech companies cannot narrow their scope to technical teams if they want to produce great software products. This is especially true when providing SaaS (Software as a Service), where customer support and training are bundled as part of the product. When the entire company is the product, it makes sense to have a technical COO who is responsible for operations as a whole.

The tech COO is responsible for overseeing product management, development, and support. She also works with other departments such as sales, finance, and HR. The COO also likely coordinates strategies and higher business goals with other management roles such as the CEO, CFO, or CPO. She is expected to translate the organizational and product vision into operational and technical vision with realizable long-term strategy and plans.

At its heart, the COO role involves taking charge of operations and the day-to-day delivery of the company's software. It also often involves working in interconnected roles alongside finance, sales, account management, marketing, PR, HR, and IT – providing there is no CTO, or if roles overlap.

While nearly any tech organization in the process of scaling up will have someone in this role, he or she is not necessarily labelled a COO. Your specific role may take on any of a variety of names inside your organization, including CTO, tech leader, co-founder, or any of a dozen other names. What's important is that you take on the role of developing company vision into operational and technical vision and strategy. For the sake of clarity, I have chosen to use COO and tech leader throughout this book when referring to this role.

My own career as COO reflects this diversity of roles and responsibilities. I began working in development at Nmbrs and grew to manage the technical teams as CTO. When we noticed the close connection between my role and other disciplines such as customer support and the close interaction with internal stakeholders such as marketing and sales, we chose to broaden it to COO/CTO to cover a larger scope within the company, including teams focused on delivering value to the customer. This gave me the opportunity to develop the full operational model from vision to value, which included product development, infrastructure operations, customer support, and startup/onboarding.

The Enterprise Architect

In my experience, the role of Tech COO is comparable to that of an architect. You design and conceptualize structures and implement them for your organization in the same way an architect designs a building. Your "blueprints" will become concrete implementations and even the smallest detail might be make or break for your organization. You have to involve stakeholders to discuss vision, implementation, and operational details which could contribute or detract from the success of the organization.

The Tech COO designs the processes, strategy, and activities which support growth and enable a company to run smoothly.

Most scale-up organizations are at a size where the COO takes on numerous roles inside the company. While these roles will naturally evolve and become separate as the company grows, organizations in early stages of growth simply don't have the resources to hire specialist managers for each task. For this reason, many COOs find themselves taking on the role of business enterprise architect. This role is natural, because it closely relates to defining an operational model, which is necessary for delivering value to customers.

Enterprise architecture is the discipline of organizing and structuring processes, managing data, implementing information systems, and integrating diverse business units or departments to enable organizations to better and more efficiently achieve their goals.

This includes the technology and applications or platforms used to extract data and information to be transformed into business knowledge. Executive teams use structured knowledge to drive the business at a higher level and maintain a solid structure for everyone contributing to the organization. As COO in a small organization, this is largely your responsibility.

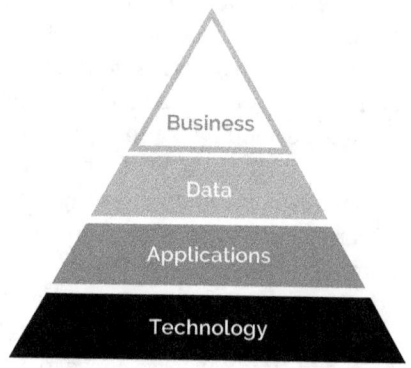

Fig.2. Enterprise architect pyramid

The enterprise architectural model is often split in 4 layers:
- **Technology:** all devices, servers, cloud platforms to enable applications.
- **Applications:** the diverse applications that enable business decision making and strategies.
- **Data:** the information structure collected by the applications and its analysis and usage.
- **Business:** the goals, objectives and strategies to drive the organization.

As COO, your role is primarily technical at the start. As your organization grows, it will require you to take on more and more creative strategy and organizational design. Scale-up is the ideal time to begin implementing this creative strategy and design, as you have more resources and more need for operations to structure work and processes. Working as COO is about more than simply using execution to handle maintenance and to keep your organization running. It's about designing the base processes and structure which will become your organization.

Your operational model affects what your organization can do, how it grows, and where it is able to grow. Much like an enterprise architect, you must design structure in a way that enables the organization to move in line with the vision. In this way, as COO, you are literally designing the future of your organization by building the support structure which will enable it to move into the future.

Cooperation between COO and CEO

If the CEO is the ship's captain, the COO is his first mate. The COO takes on the details of day-to-day operations, ensuring that the organization is moving, and in the right direction. For this reason, the COO and CEO roles are naturally closely linked.

While the CEO will often make the big decisions, the COO is responsible for bringing those decisions to life. A close connection and shared trust are vital. Both must share vision and align on strategies, because the COO is responsible for moving the organization in the direction set by the CEO.

Complete trust between the roles is also a necessary part of building a functioning team. The CEO must trust the COO with autonomy and independence so you can do your job well. The COO must be able to use their own authority to make decisions without undermining the CEO's strategies or vision.

Organizations including Google and Microsoft have vastly benefited from having COOs to support and guide their CEOs and vision. Bill Gates was guided by COO Jon Shirley, who was largely responsible for Microsoft's financial and managerial structure and is largely credited with guiding the company through its rapid growth in the 1980s.

Google's Eric Schmidt functioned in a similar role, building corporate infrastructure to maintain quality and development speed throughout Google's rapid growth. Both take on the role of "sidekick", or COO supporting the CEOs vision.

In short, the CEO owns the company vision. Good communication and a constant feedback loop will help you to align progress and direction. Over time, this will help to avoid misunderstandings and mistrust, which will negatively impact the whole organization. You, as the COO, own operations, which must be aligned with the company vision.

It's a good practice to clearly define the roles and responsibilities of the CEO and COO from the start. This will give each the autonomy to make decisions directly without the need of involving both parties and without 'stepping on each other's toes'.

Aligning in this way will ensure that employees receive a consistent vision and strategy from both the CEO and COO, with no conflict of decision issues when consulting either. Aligning goals adds value to the COO role by preventing the CEO from undermining the role. This is essential in a tech organization, where the COO directly aids the CEO with knowledge and experience in technical and product development scopes.

Both roles need a very close connection and strong relationship to make things work. Each has its purpose and, with proper alignment, they will deliver great value to the organization together.

Connection with HR and Finance

Working closely with other departments inside the organization is a crucial element of performing the role of COO. In many cases, your role will encompass elements from many different departments. Your role will logically begin with those connections, but it is crucial that you retain those connections as your role evolves. Tech leaders require very close connections with many areas of the organization, but HR and Finance are among the most relevant.

One of your primary goals as a COO is building strong teams inside the organization. This requires very close cooperation with HR, where you assist with guidance, profiles for recruitment, candidate skills, and culture-fit.

Working to define operational processes and roles with and for HR gives them the tools to select better candidates, because they will better understand what you need and how people should grow within the company.

The COO role also closely involves finance. Creating and implementing operational processes and structure requires aligning budgets and costs with the finance team. This will help you determine what is possible within the budget. Most COO projects are long-term company investments relating to structure, organization, and operations which will require significant investment. A good financial overview gives you the tools to plan implementation strategically and to choose the best possibilities and timeline within the budget. You need a close cooperation with finance to achieve this.

Your role will naturally align with many other elements of the organization. Working closely with other departments will give you the tools to improve the efficiency and quality of your results. Imagine implementing a new customer support service, involving hiring, tooling, and training. You would need close cooperation with HR and Finance to optimize your success.

No matter what your job title in the company, you will work to create the infrastructure and operational processes necessary to enable growth and long-term success. Understanding your role and defining how you and other players interact in it, as well as your responsibilities and goals in overseeing product development, will enable you to set the right focus and goals as you move your company towards growth.

In conclusion

As COO or tech leader, you take on the responsibility of translating organizational vision into technical operations. This requires a deep understanding of where your organization is headed and why so that you can develop operational structure to take it there. Your role impacts every part of the organization and its growth, which will give you the opportunity to develop and design a company that meets the needs of employees, brings value to the customer, and delivers on the company vision.

Working to build sustainable operations that provide support and structure for your organization will enable it to grow. However, any solution you implement should be iterative. Tools and processes will always change, they must change. Your organization is scaling, the environment around it will change, so what was working before may not be efficient and may not even be relevant moving forward. What's important is that you have the solid foundation of processes on which to build your future growth and decision making. Designing frameworks around end-value will enable you to keep those tools relevant and useful, even when this means replacing them.

Working with an organization in the early stages of growth gives you the opportunity to truly design infrastructure in a way that supports the capabilities and value you want to deliver. Building that requires an operational framework.

PART II

THE VISION TO
VALUE FRAMEWORK

The Vision to Value Framework covers the basic foundations of operations inside a tech organization, based on my experience scaling my own company.

Understanding the framework, its pillars, and the factors affecting it will be crucial to your ability to adopt and use it yourself. Scaling your organization will require significant organizational and operational structure, which, in a tech company, differs greatly from operations in companies producing physical products.

Introducing
the Framework

"A company can seize extraordinary opportunities only if it is very good at the ordinary operations"
– Marcel Telles

Working at Nmbrs, we often had to design our operational strategy from the ground up. As we experimented and used research to develop new and better operational strategies, I noticed most of the concepts and problems I worked with fell into 5 categories; vision, work, processes, people and teams, and data.

Over time, I defined these categories into a recognizable and structured model, which we could use again and again for repeatable success. I call these 5 categories the operational pillars, and they make up the tenants of my Vision to Value Framework. These pillars affect most aspects of normal operations management including short-term day-to-day work and long-term planning, vision, and strategy.

Here, the vision encompasses "why the organization exists", work translates to "what the organization does", processes outline "how this work is done", people and teams defines "who does this work" and data brings everything together.

These pillars are broader than those normally applied using Agile frameworks. They create a more holistic and comprehensive view of what goes into making a tech company run. Rather than focusing on work, they deliver insight into high level strategy that determines not just what you do but how you work as a company. Approaching your operations and strategy in this way gives you a better perspective, so that you know where you want to take your organization and can develop the strategy to get there.

Some background on operations

Operations management is the process of designing and controlling the production of goods and services. Its goal is to ensure business processes require as few resources as possible, are as effective as possible, and comply with or exceed the expectations of customers.

Operations has a broad scope, including everything from process inputs such as raw materials and ingredients to final output, such as manufactured products and services, as well as day-to-day activities resulting in the production of a product or service. Here, processes include all linked tasks and activities. These tasks work together to accomplish an organizational goal, such as software development and code contributing to a software product, or customer service and marketing working to drive sales. Managing these processes is an important part of operations and is typically defined as business process management (BPM).

In most organizations, the COO oversees this process, creating and setting up strategy, which eventually has an enormous impact on business success. This concept has evolved a great deal over time, but dates to ancient times. Ancient Egyptians used operational planning to build complex structures in inhospitable environments. Guilds during the Middle Ages began to perfect operations as they worked to move goods across trade routes. However, operations management as we know it today truly came into being during the industrial revolution with the invention of mass production and factory lines, as popularized by Henry Ford – who used strict process management to streamline his production and reduce costs.

No matter what your business or what you produce, operations ties into every aspect of production. In modern, tech-based companies, software intrinsically changes operations inside the company. Here, input is realized through the product vision, internal stakeholder requests, and customer demand. Output is normally realized in the form of a software product, which is deployed online or distributed to customers.

The process of designing and developing software aligns with production, which is the process of acquiring raw materials (in this case a concept and teams to produce it) and developing a final product. For this reason, operations management in tech companies normally revolves around software development cycles, making it substantially different from operations revolving around acquiring and processing ingredients or raw materials.

In addition, there are many other relevant processes and structures, such as customer support, customer on-boarding, sales, finance, and HR. Each of these processes, including their relevant teams and departments, tie in to deliver a quality product and service to customers. As COO, you must define strategies, process design, team structure, quality frameworks, and improvement frameworks for each of them.

The Vision to Value Framework

The Vision to Value Framework encompasses operational structure and quality delivery. It creates a cycle or loop connecting the actions, processes, and structures needed to develop an organization capable of delivering a quality software product as you scale.

How does this loop work? The act of delivering a product generates data and information needed to further refine operations. This creates a cycle, where the two feed into each other.

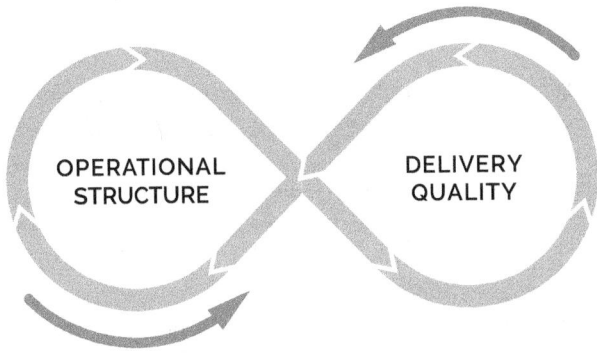

Fig.3. The Vision to Value Framework Loop

Operational Structure – In the Vision to Value Framework, operational structure is organized around 5 pillars: Vision and Strategy, Work Management, Processes, Team & People, and Data & Information.

Delivery Quality – This encompasses how well you can deliver a product to end-users. Delivery quality includes 7 qualities or traits, including product quality, service quality, development velocity, team motivation, innovation, cost-effectiveness, and compliance.

Designing operations with the 5 Pillars in mind will naturally improve delivery quality. However, it's unlikely that you can improve each of the 7 qualities. Instead, you will have to observe your organization, stakeholders, customers, and users to find improvement points that relate to your organization's goals, so you can improve the aspects of delivery quality that matter most.

This process becomes an infinite loop because the environment is always changing. Your organization is dynamic and growing, your customers and stakeholders have changing needs, your competitors launch new products and services, new technologies arise, and so on. Operations management is a continuous process of building structure, receiving feedback, and improving that structure to meet your goals as they change. The Vision to Value Framework gives you a basis to build that structure on. You have to continue to update your operational model around it, so that it continues to meet your organization's needs.

Delivery Qualities

Operations management is the process of creating structure, including plans and strategies, to manage the challenges of working in that organization's environment. It's necessary to structure and invest in operations. At the same time, the only value of operations is their result. If you are striving to build operations, you are striving towards an end-goal of quality, of value for the customer, and organizational efficiency.

With that in mind, I like to view operations design not as a series of processes to create operations but to create a result. For most organizations, that end goal will look something like this:

"Our highly motivated teams continuously deliver an innovative product that customers love. This delivery is managed on-time, within budget, and within regulations".

If operations are not about end-result, why bother? Here, I like to use an analogy about gardening. You won't just wake up one day and find your garden full of plants bearing fruits and vegetables. You have to choose seeds, plant them, and water them so they eventually grow.

Operations can be compared to feeding and caring for those plants. The things you take care of, invest into, and put time into are the things that grow. This also highlights one of the hardest and most complex concepts in operations. You have to be patient. Even if you have everything you need to create a good result, you still have to take care of it and wait for results to appear.

Building your operations according to the Vision to Value Framework's 5 pillars will eventually result in 7 delivery qualities in your organization. You can create the conditions for them, you can build good teams and processes, but you must wait to see quality results.

Product Quality – Operations should create a structure to manage product quality so that the organization outputs a product that customers like. The product should be quality enough that it is easy to sell. In addition, operations must define what quality is, such as a low number of bugs, working features, performance, or ease-of-use.

Service Quality – Standards should be in place to manage and maintain service quality. Customers should be happy with the product, with response times, and with how customer service representatives respond to them. Here, operations should be concerned with internal escalations, monitoring, and with organizing teams in a way that allows customer service to receive information needed to remain helpful.

Development Velocity – The product must evolve at a pace that meets customer and stakeholder expectations. Processes must be in place to invest in growth and innovation while maintaining quality and continuing to meet the needs and expectations of customers now.

Team Motivation – Teams should have the structure and support to feel happy, be empowered, and to take initiatives – without stepping outside of their roles or responsibilities.

Innovation – Structure must make room to try new things and to ensure experiments are successful so that they make the product better. Your business cannot continue to deliver the same things in the same technology. You need structure that supports continuous innovation in technology and in your product.

Cost-Effective – Your operations should be structured so that development and running costs fit into the organization's budget. This means taking all costs into account such as personnel, servers, internal development, quality control, automation, etc., and creating a structure that works while remaining inside budget.

Compliant – Solutions, including work, the product, policies, and procedures, must be within local regulations such as the GDPR and compliance standards like ISO and ISAE.

While your exact operational goals will vary, they will always work inside these parameters. The Vision to Value Framework brings strategy, work management, process management, people and structure together with data and information management to achieve these 7 qualities.

Each of these qualities is separate, but, like the 5 pillars, each impact the other. They are results, but when you try to cater to or improve a specific quality, it's very likely that other qualities will suffer. In fact, some directly compete.

For example, if you invest or model your operations around cost-effectiveness, it's highly possible that your product or service quality will decrease. This is especially true for organizations with limited resources, which is why I discuss the idea of choosing your primary market strategy later in this book. If you have unlimited resources, you can "max out" all 7 delivery qualities. However, the chances of you having these resources are close to zero. As a startup or organization in the early stages of scaling up, it's incredibly important that you choose one. You can typically do so by setting priorities.

If you have investors, cost-effectiveness is less relevant and, if you can afford it, investing more will typically generate better results and higher product quality. You can then use that to grow your business.

Often, choosing delivery qualities is less about selecting one and sticking to it forever but rather about choosing which qualities to invest in first based on how they impact your business and your goals.

I prefer to prioritize customer-related qualities such as product quality and service quality. The others, which are primarily internal concerns like customer service, compliance, and innovation, only have a direct impact over the long-term. My organization needed product quality and service quality from the start, so we chose to invest in those.

In any instance, these 7 delivery qualities are the result of tweaking the 5 pillars. You can always expand or improve one quality, but often at the expense of others. However, once you learn how to push the right buttons and move your operations in the right way, you can control these qualities to design your organization in a way that meets your strategy and goals.

Operational Structure

As COO, your goal is to create organization and operational structure to support delivering value to customers. You have to build the structure that motivates teams to work on the right things, in the right way – delivering the value and quality product your customers want or need. You must take details into account, even when they go well above and beyond what Agile frameworks encompass, which is where the Vision to Value Framework comes into play.

The Vision to Value Framework uses 5 pillars, which define important aspects of operations that you must take into account.

1. Strategy (Why)
2. Work management (What)
3. Process management (How)
4. People and Structure (Who)
5. Data and information management (Bringing the pillars together)

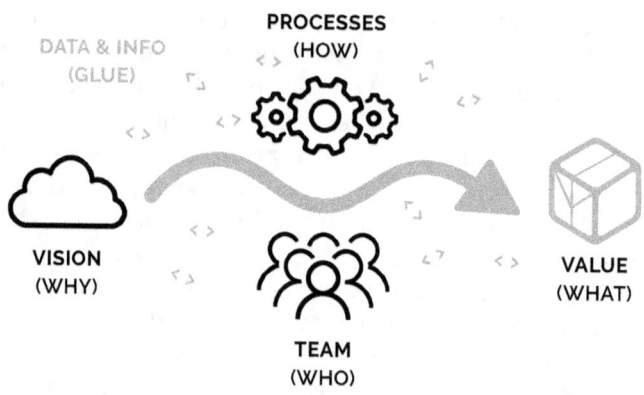

Fig.4. Operational Model

These pillars embrace the full operational model, spanning strategy and vision, work management, data management, teams, and much more. Each is interconnected, but also has its own strategy and purpose.

Designing a successful operational model requires balancing each of the operational pillars to create a single system that feeds into itself, so that each pillar supports the rest of the operation.

Vision and Strategy

Your strategic plan is the heart of your business. It defines not just what you do but why you do it. This allows you to align your organization at a high level with clear goals and strategies.

Defining the "why" for your business gives you the opportunity to define high-level strategy. This ensures that everyone in operations understands company vision and what your goals are.

This mutual understanding – where founders and managers set long-term goals and business priorities – is crucial to avoiding miscommunication and misunderstanding, which will hurt the business in the long-term.

The strategic plan includes vision statements, cultural core values, mission statements, ambitions, business plans with market and SWOT analysis, strategic themes, and objectives. If this part of operations is clear and visible, everyone in the company knows what the company is aiming for. You may not yet know how to get there, but you know where you want to go and why.

Depending on your organization and which phase of growth you are in, you may or may not already own these assets. However, your strategic plan is essential in supporting your operational model and the organization as a whole.

Work management

Work management is a crucial aspect of operations. It is also one that affects both day-to-day activities and long-term strategy. Here, you must define activities contributing to delivering value for the customer and to delivering company vision. These chains of activities are commonly known as value streams. They break a vision down into actionable activities and work items, so that people can achieve them.

Defining activities for your organization, the deliverables for those activities, and how those deliverables create internal or external value will give you a way to define and measure work inside your organization. This will, in turn, enable you to plan and prioritize work, communicate and measure progress, and align everyone with those activities.

Process Management

Process management involves defining how you want to perform work. This involves creating concrete roles, responsibilities, and processes. It also means defining interactions and collaboration of these assets. Finally, process management involves defining tooling and how to map and model processes in tooling.

Process and output monitoring are important factors of process management. Why? You have to be able to track progress and improve as part of your process. This is especially crucial in scale-up environments, where the environment and required deliverables are constantly changing. A process that worked smoothly 6 months ago may not be effective in your new environment. Monitoring your processes ensures that you can update as requirements change with no gap in productivity or output.

Clearly defining how you want work to be done helps people to understand how their work connects to value and the strategy, how it connects to the responsibilities of everyone else in their and other teams, and ensures that work is completed in a way that aligns with company values.

People and structure

This pillar of the operational model – people and structure – is about team management. Good team management requires that you have a defined purpose, scope, and strategy in place for your teams. Team performance management is about organizing your teams, roles within teams, and how they collaborate with each other to align with strategies and objectives – typically using a combination of operational frameworks such as Agile and HR frameworks like Belbin. This is essential for long-term success, especially as your organization begins to grow and tasks are spread across teams, which must align to create a working finished product.

Team performance management enables you to align everyone with long-term value-focused deliverables rather than specific technical goals which may not deliver value to the customer.

Data and Information Management

Data and information management is the figurative glue holding your operational model together. It functions to connect and foster communication and quality management between every other pillar of the framework. Here, you use data from every other pillar to enable other pillars, support functions, and to share progress and communication with teams and stakeholders. Data also enables you to track progress to ensure goals are met, value is realized, and work is performed in a way that aligns with values and targets.

Sharing data in the right way impacts your organization at every level, from internal process management and the platforms you use to document and share information to team performance and alignment between teams. The processes you use to make work visible to stakeholders affect your organization at a strategic level, because they impact stakeholder's view of what you are achieving and why.

Each of these 5 pillars is intrinsically connected. If you are setting up or improving technical operations, you will have to take all 5 into account. If something is missing, your operations will not work correctly. However, it's also important to keep in mind that this is a conceptual framework. It's not a fill-in-the-blanks solution. Instead, it functions as a set of guidelines which you can and should adapt and modify to meet your own organizational needs.

Vision and Strategic Plan

"Be stubborn on vision but flexible on details."
— Jeff Bezos, Amazon

Vision and Strategy is the pillar in the operational model where you define why you want to deliver value and to whom.

Your vision and strategic plan give your organization the foundations to deliver value to the customer. Without them, you cannot set up the functions or goals to create a sustainable company. In most cases, company vision comes from the CEO, co-founders, or a series of stakeholders, who each have their own input which must be aligned to create a vision. As COO, you work with stakeholders to describe and share the strategic plan and to define how the company will move from that vision to the end-goal of delivering value to the customer.

This strategic plan must be high level, connecting to strategy, company vision, as well as to real work and value streams. It's the point where you connect where you want to be as a company (vision) to your end-goal (delivering value).

Everything begins with vision

Any organization must exist for a reason. No businesses will have a starting point with no reason for its existence. Any founding CEO or team will naturally have goals, a vision of what they want to produce or create, or a problem they would like to solve. These goals are abstracted at a high level to form vision statements to guide the organization, mission statements to guide strategy, and product vision, to steer product development.

Vision statements are simple statements defining the overall purpose of the organization. Amazon wanted to make books cheaper and more accessible. Facebook wanted to connect people. Defining your vision means summing up the primary goal of the company and its reason to exist, which will help you to better define your end-goal and how you deliver value to the customer.

This normally translates into a long-term vision statement, like Microsoft's *"To help individuals and businesses realize their full potential"*, IKEA's, *"To create a better life for the many people,"* or Disney's, *"To make people happy"*.

These vision statements can sometimes feel unrealistic and utopian, but why not? A powerful vision statement will help your organization to understand its purpose and goals for every activity, behavior, and decision.

Your long-term vision helps you align teams and individuals with the company's purpose and enables leaders throughout the company to define their own strategies with their own scope inside the company vision. This also means all strategies, goals, and therefore all work, align with the vision statement.

If vision defines your organization's purpose, the mission statement describes how you achieve that purpose. It's typically more detailed, and in some cases, will refer to specific products and goals.

Here's an example of Apple's mission statement:

"Apple designs Macs, the best personal computers in the world, along with OS X, iLife, iWork and professional software. Apple leads the digital music revolution with its iPods and iTunes online store. Apple has reinvented the mobile phone with its revolutionary iPhone and App store, and is defining the future of mobile media and computing devices with iPad."

Both the vision and mission statements should be powerful messages which you use to align your organization and your goals.

> **Strategy Tip:** Define simple but clear vision and mission statements everyone can remember and align with.

Product vision forms the backbone of your strategies and decisions because it answers key questions about your company. For example, do you have to take international expansion into account to meet customer demand? What kind of customer support will you need? What are your biggest challenges? Your product vision answers these questions by defining where you want to go.

It's also crucial that you have a deep understanding of your products and why you are developing them before you begin to define the operational model. Why? The operational model supports product development while enabling quality and value assessments for those products. Understanding the product vision and purpose is therefore essential to developing an operational model that can achieve its goals. Here, your product vision should answer questions like "who is the product for?"

Your organization, structure and strategies will change dramatically depending on whether your organization is business (B2B), or consumer (B2C) oriented.

Business oriented products are mostly connected to existing business processes or systems of record. They tend to focus on strong relationships with customers, which can sometimes manifest as partnerships. Here, your product vision must enable you to create an organizational model where teams grow in relation to the number of customers - enabling you to keep growth in pace with products that rely on difficult customer acquisition.

On the other hand, consumer facing products are typically luxury products that people like but don't necessarily need. Here, your ideal goal is to create a new demand that people depend on. For example, Facebook relates to basic human needs like communication, sharing, and belonging; which quickly promoted it into a product people don't want to live without. If your product is a "nice to have", your product vision would result in creating a strategy based on user analysis and retention to promote growth by creating added value and improving lives.

Understanding how your product works, what problems it is solving, and how it is solving those problems often means approaching your product as a user. It's also a good idea to question why people bother to use your products. For example, does your product help them optimize processes, is your product creating a new demand, or is it a luxury product?

You should have a good understanding of how your product differs from your competitors. Is it better? Cheaper? More user friendly? Easier to use? Does it offer more or fewer features? Better performance?

No matter what products you are developing, it's important to understand why consumers purchase or abandon it. Your internal services and teams will be organized around the customer's demand and delivering value, which necessitates developing monitoring and quality metrics around the product vision.

Understanding what your product contributes and why people buy it will help you in creating guidelines for quality frameworks and for defining value streams. You also have to understand who your end-users are, which will affect how and why they choose products and even what value is. This product vision will then help you create structures to align teams towards developing products that meet goals and deliver value to the end-user.

Strategy Tip: Integrate product vision into operations and strategy.

A perspective on vision

Having a well-defined company and product vision helps you set clear goals because you understand your targets and the real value you are delivering to the consumer. Once that vision is clearly defined, you still have to translate it to direct and actionable work. If daily work and activities inside your organization do not relate to the vision, the company is not moving in the direction you have set.

Creating this connection between daily work and vision means ensuring that everyone within the organization understands the vision. Developing a clear structure connecting vision to work is important, because it enables individual contributors to work towards that goal with every decision and output.

Fig.5. Connecting Vision to Work

Your company vision represents the why or purpose of your organization. Your strategy is a high-level path describing how to get there. Once you have that path, you need to walk it. Taking this step is crucial to actually achieving goals, but it is one that many scaling organizations struggle with.

For example, when you work with Agile, your daily work is mapped to epics and stories. These epics and stories must be connected to company goals, or they won't help you move towards your long-term strategy. If you're just doing work with no focus or milestones to support company goals, it's likely just maintenance. You're doing what you need to "keep the lights on", but most likely not what your organization needs to move forward.

You have to define a structure that links daily work to goals, strategy, and vision. Your structure is very dependent on the tools and frameworks you use, but it creates a clear end-value by enabling stakeholders and teams to track how and when work contributes to goals. This will aid in aligning the organization, communication, and when making decisions in product councils and strategy sessions.

OGSM is one strategic planning process that can help you achieve this. OGSM links long-term vision and strategy to short/mid-term goals, actions, and processes, so that you can measure progress, maintain focus, and share the value of actions. The concept was originally used in Japanese production companies, and was brought to the U.S. in the 1950s, where it saw rapid adoption. OGSM breaks every initiative into: Objectives, Goals, Strategies, and Measures.

Objectives – Objectives link an action to organizational strategy. What does the organization want to achieve? This should be in-line with the strategy, organizational goals, or specific business objectives already outlined in the business plan.

Goals – Goals allow you to set specific, measurable goals for your objectives. What does success look like for these objectives? Most organizations are happy with 3-5 goals. These should be specific, time-dependent, achievable, and measurable. SMART (Specific, Measurable, Attainable, Relevant, and Timebound) is one way to ensure your goals stay on track.

Strategies – Strategies define how you will achieve your goals. These should include specific actions you will take in order to attain goals. These strategies should be validated before implementation.

Measures – Measures outline how you intend to measure success or lack of it. Action planning, scorecards, and quantitative measuring are some tactics you can use to measure progress towards a goal.

OGSM is one process enabling you to connect daily work to larger organizational vision and strategy. It's valuable because it incorporates measuring success as part of the process. However, it's not the only process available and you can choose another or create your own depending on your organization and its needs.

> **Strategy Tip:** Define how you will connect company vision to daily work and set up tools to support it.

Market strategy

Sometimes, one company may dominate a marketplace or niche, crushing the competition in what seems to be effortless achievement. Why are companies like Airbnb or Uber extremely successful in their niches while competitors are not? While many companies do so well that it can seem as though they have a secret weapon, the reasons behind their success are not always immediately obvious to onlookers.

However, there is a formula behind every type of success, and in this case, the formula is often using competitive strategy. Here, organizations choose a strategic focus and keep it consistent throughout their lifecycle and operations so that they can consistently deliver to meet a specific customer demand.

Organizations need to embrace a competitive strategy to succeed in their marketplace. According to Fred Wiersema and Michael Tracy[6], most organizations do so using one of three primary strategies that define how companies focus their operations: Product leadership, Operational excellence, and Customer intimacy. Focusing your product and company strategy on one of these options will allow you to define your competitive goals, which will help you to target your specific customers and realize more direct value to your demographic.

Why not apply all three strategies and be the best of the best? These strategies conflict with each other when fully implemented, so making a choice is the only option if you want to truly excel at any of them.

Product Leadership: Product leadership companies specialize in creating and delivering superior products and services at a premium price – typically with an added quality, feature, or service value over their competitors. This type of company focuses on staying a step ahead with product innovation and continuous development. This is a common strategy for SaaS companies and

---

6   https://www.amazon.com/Discipline-Market-Leaders-Customers-Dominate/dp/0201407191

cloud-based products, where a single product is delivered to all customers with no options for customization or personalization. Apple, Sony, and BMW are good examples of companies adopting a product leadership approach, and they are best known for quality or cutting-edge products rather than for low prices.

Operational excellence: Companies adopting an operational excellence approach offer value to the customer by using technology such as automation and process optimization to reduce internal operational costs and offer lower prices to the consumer. This model is suitable for any market where product is not a differentiator and where competition is fierce. Retail companies such as Amazon, Walmart, IKEA, and McDonald's often adopt this strategy, because the deployment pipeline and supply-chain management can be a strong differentiator to cut costs.

Customer intimacy: Companies focusing on customer intimacy work to offer personalized products and services to meet individual customer needs. These organizations treat customer relationships like partnerships, developing a strong understanding of the customer's process and problems to deliver custom solutions. This strategy requires that you work with specialized partners to achieve the levels of personal and custom service consumers expect. IBM, Home Depot, and Nordstrom use this strategy.

While each of these strategies can offer some value to most companies, it's important that you choose one and only one so you can excel at it. You can still put some focus on the remaining two strategies, but you cannot be the best because there will always be factors from one strategy which conflict with factors from your primary strategy. Trying to achieve everything will damage your ability to excel at anything, which will harm your organizational results.

For example, organizations choosing to focus on product leadership must ensure that their product is better than the competition. It's theoretically impossible to be both the best and cheapest, because there are trade-offs and compromises that will result in losing both strategic objectives.

How can you choose your competitive strategy? It should be closely tied with your company vision and agreed on by the full leadership team. Most importantly, it depends on your organization and it will work to define your organization and what kind of company you want to be. This choice should not be taken lightly, but you can use your existing ambitions, culture, and vision to make it.

As the COO, you are responsible for ensuring that the strategy is identified and that the operational model is designed around it.

How can you do so? Your approach will naturally depend on the strategy your organization has chosen.

For example, if you choose product leadership, your goals must focus on creating a strong structure around product development with core and R&D teams to continuously develop new products or features. Here, your mindset and operations should surround creativity and innovation. You should focus on business and quality metrics for product engagement as well as product metrics that differentiate your product to succeed with this strategy.

If you've chosen to focus on operational excellence, your operational model must ensure that core processes are designed to optimize and reduce waste. Your teams need to have a mindset of efficiency and waste reduction. You will have to develop structures that identify inefficiencies through all activities so that you can reduce operational costs over time. Here, you may also set up internal teams with no focus other than automation and process improvement. Your most important quality metrics include process efficiency, time to deliver, and costs.

If your organization chooses customer intimacy, your goals will surround developing a solid network with specialized partners so you can offer solutions that adapt to your customers' needs. Teams should be organized around customers or customer segments, enabling them to offer better individual support. You should focus on how you and the organization can better understand customers to deliver more value. Your most important quality metric is customer satisfaction.

Strategy Tip: Identify your organization's sole competitive strategy and design your operational model around it.

Understand the provision model

Product delivery, or how you deliver your product to the customer, will greatly impact how you structure operations. It should be taken into account when creating strategy and making decisions for your operations. For example, are you delivering a SaaS product where users can sign up online and immediately access a free trial? Or, are you selling a product which you must manually provision to each new customer?

Your provision model greatly affects your operational strategy by determining how onboarding should be handled. If you're using an automated provisioning model, where customers sign up and immediately have access to your product, it's extremely important to have a proper monitoring system and metrics in place to measure performance. At the same time, automating the onboarding process doesn't necessarily mean you won't have human contact with your users.

On the other hand, if you're offering complex products and services with a complicated migration before product use, it must be covered by a specialized team to onboard new customers.

This is important for your organizational structure, because you must set up processes in line with your competitive strategy. You might have to create a startup team to onboard customers, a product team to ensure proper delivery, and so on. Each of these teams must align with company strategy and operations.

Creating the proper internal services and automation to follow up your onboarding process – whether that's a recurring subscription business model, a monthly fee, or a one-time purchase – will help you to structure your teams in a way that supports the provision model. These are relevant questions when you are designing technical operations to cover risks, ensure business growth, and offer value to the customer.

Strategy Tip: Align your provision model with vision and strategy.

The (Business) Plan

Understanding your strategy, vision, and cultural values is an important part of the scale-up journey. However, you still have to define these ideas and strategies in a way that can be shared, worked out, and used to measure and track progress. Your business plan describes the ideas and strategies behind your organization and its vision. This allows it to function as a guideline which defines what you want to achieve, how you want to achieve it, and how you will measure your progress. While these factors can and should change at any time, you need the guidance of a business plan so that you can track progress and determine if you are meeting your goals.

While less important in a small startup, the business plan becomes increasingly important as you begin to scale up. It's crucial to have a clear picture of where your organization wants to go and what that journey should look like. Why? I've seen many companies, small and large, where founders had a clear picture of where they wanted the organization to go, but, because they didn't share that plan across the organization, it was never clear for everyone else. If most of your organization is performing work without a structured goal, that work won't likely align with your vision and strategy.

Business plans are commonly used to inform potential investors and other external parties. Defining how the business is constructed – growth projections, people, sales, costs, etc. – is important for leveraging stakeholders and gaining buy-in for funding. While the value is most apparent for external stakeholders, I strongly believe in the importance of sharing the business plan with employees at every level of the organization.

If your employees are unaware of your plan, your organization is akin to an aimlessly drifting ship. People, even those working tirelessly, will not move the organization towards its goals if they are unaware of those goals. Most importantly, sharing the business plan and working to facilitate internal buy-in will encourage everyone to truly care about where the organization is headed.

Your business plan will also function to help you gain buy-in from stakeholders and investors. It's valuable for helping members of the management team make strategic decisions. For example, if you're moving and need a bigger office, how much bigger? If your business plan is done well, and updated as the market changes, you could easily check growth projections to see how much space you'd need for the coming 5 years according to projected growth.

Your business plan is not a static thing which you can create once and leave alone. It is a thing that is as vibrant and changing as your business and must be frequently updated to reflect the present reality and projections for your company based on the current market. It's also a good idea to maintain a copy or historical record of changes to see how it, and your organization, evolves over time.

> **Strategy Tip:** Develop a clear and strong business plan which you can share with stakeholders to gain buy-in as well as with employees to create a culture working towards the business objective.

Involving teams in strategy

Once you've identified your primary competitive strategy, it's important to translate it to individual strategies for different scopes and teams. For example, product, tech, Quality Assurance (QA), support, sales, etc., will each require their own operational strategies to succeed. Taking the time to define these channels will enable the business plan and help you to define scopes for individual teams.

This also means involving your teams and their leaders to ensure you can develop a strategy that makes sense for individual team capabilities and creatives. Involving teams in working out how they will achieve company vision also works to motivate individuals towards achieving that vision because they naturally become a part of it.

You have to actively discuss capabilities, expectations, and outputs with teams. Here, you should work out how to achieve company vision and why it is important, and how the scope of any given team fits into that larger puzzle. Encouraging teams to look beyond the scope of their individual role and towards that of the organization as a whole – including how they contribute – enables everyone to collectively contribute to a strategy which will guide day-to-day actions for the near future.

> **Strategy Tip:** Work with your specialist teams to develop strategic themes.

Set focus with Goals

It's important to understand the scope of your vision and long-term goals, but also to translate your vision into achievable and attainable work. For example, when you work with several teams, it can be difficult to ensure they align and work in the same direction or with the same priorities. Setting goals gives you the opportunity to focus on specific and achievable units of work, so you can steer teams, stakeholders, and strategy towards that achievement.

While goal setting might seem simple, I have to say, it is not. It's not easy to formulate goals in a way that describes what you want to achieve and do so in a way that people can translate into actions and deliverables. OKR is a good framework to facilitate this process by describing goals in the form of a main objective and its key results.

OKRs consist of 3-5 high-level objectives, each with 3-4 measurable results. These can be defined at an organizational, team, or individual level and could be incorporated at every level across the organization. OKRs were first used in the 1970s at Intel, where Andry Grove kickstarted their adoption in one of the largest tech companies at the time. Since then, OKRs have spread to thousands of tech companies, especially following Google's early adoption of the method – leading to their integration in companies ranging from LinkedIn to Zynga to Oracle and Twitter.

Objectives – Here you define 3-5 achievable goals. These should be bound by time with actionable steps to achieve results, but should be ambitious. For example, "Improve document and information management procedures" would be a good objective if your aim is to make data more available across the organization.

Key Results - Define 3-4 measurable results under each objective. These results should be quantifiable, measurable, and achievable. They should be difficult but should contribute to objectives (such as growth, performance, revenue, customer satisfaction, etc.). If you were to define key results contributing to the objective listed above, you might use items such as, "implement software solutions for tracking incoming documents", and "review and implement solutions for tracking incoming requests".

Using OKRs means setting objectives in line with high-level strategy at the level you are working with. In most cases, OKRs are part of quarterly planning and progress review, but may also be set annually or monthly. Once defined, they give you a clear way to track what you want to achieve, as well as the key results which stakeholders need to know to see if those objectives are being reached.

Once people and teams begin working, they can update results on a weekly or monthly basis to keep stakeholders informed. In most cases, objectives should be ambitious enough that achieving 70-75% of results is considered a success.

OKRs are a tool you can use to translate strategy and vision into definable and measurable goals. However, it's important to continue to review OKRs and update them as needed because they are a tool and any tool has to fit what your business and your people need now.

It's relevant to understand the size of your goals and why you are setting that size. Goals that are too small can feel like a normal work task or an epic and those that are too big can feel too far away to keep the team focused. It's always a good practice to set goals within 3-6 months so that they are more immediate and achievable.

You should also consider setting goals at different levels, such as organizational, team, or any other layers you may have in between. Here, you can nest goals so that you align individual and team goals with higher-level achievements. This is a very useful practice when tracking how and where teams are contributing to an organizational level goal.

Fig.6. Cascading goals

Nesting goals stimulates bottom-up initiatives, where teams create their own goals based on high-level goals. This is especially useful in flat organizations where team autonomy is necessary for success.

You'll have to consider how often your organization sets goals and how quickly you measure or expect results. Here, periodicity is determinant for setting the pace of your organization as well as for gaining buy-in with teams.

For example, many organizations will set a batch of quarterly, semesterly, or yearly goals and spread (or throw) them throughout the organization. Unfortunately, this approach means that teams aren't really involved with formulating goals. In most cases, goals are also set without really determining which resources will be used to reach them or what capacity is available to work on them. This results in confusion rather than focus, because it can be difficult to focus on a goal when it is either not well-connected to resources and outputs or when there are several goals for the same team.

Goal prioritization is another consideration. If you set too many goals, it's impossible to develop everything at capacity. If you can prioritize goals based on factors such as importance to business value, you can allocate capacity to the highest priority and reallocate as goals are completed. If you don't start goals until you have the necessary resources, capacity, and focus, your ability to manage those goals will be much higher. This requires using a backlog.

This backlog should be created as a joint effort between management and teams and should align with the business plan and strategies.

Here, each goal should be formulated with a clear objective and key results and with enough information to define which teams are included in its scope. This enables you to assess capacity and planning so that you can prioritize and launch goals as necessary resources and people are free to focus on them.

Setting goal deadlines is often difficult because it's sometimes impossible to determine how much work is necessary to achieve results. However, you can always set deadlines based on external factors such as seasons, important events such as product launches, and so on. While you may be tempted to set arbitrary deadlines, it's important to set deadlines based on clear reasons, so that teams can align and understand why they are there.

Goals help you increase team-level engagement for high-level objectives by setting a clear focus and direction and helping individuals to connect daily work to long-term vision.

> **Strategy Tip:** Only launch company goals when you have the defined resources such as a timeline, capacity, and resources to achieve them.

Conclusion

Once you have a clear understanding of the organization's vision, business model, and product vision, you can begin creating your own technical operational vision. This will be your guideline for creating your strategies and implementing your value stream. It should describe your core values, mindset, and high-level challenges.

Your technical operational vision will define your organization's style, culture, and will be shared across your teams. It should have a defined scope or area of focus (full organization or a specific functional area), timeframe (typically 2-5 years), and company vision. Here, you can begin to define where you want to be and what you want to achieve inside the organization.

Vision and strategy reflect where you want to be while goals are real milestones which you can focus on and achieve in the near or immediate future. You can use them to generate work items and to focus on specific achievements such as moving into a country or market, using support strategy, etc., No matter what your goals, they are always specific and achievable, even if the timeframe for achievement is 2-5 years.

Finally, you must check operational vision against organizational vision to ensure there are no conflicts and that each contribute to the other.

Creating a shareable and relatable version of your vision is crucial when scaling up because individual teams and decision makers will rely on it when formulating their own strategies and decisions. Sharing your operational vision ensures that everyone has the same mindset and is aligned with the same long-term goals and strategies. A strong, present, and visible operational vision is the only way to scale your organization while maintaining focus and alignment on operational goals.

Work Management

"Doing the right things is more important than doing things right"
– Peter Drucker

Work management is the scope in the operational model where you define what. This includes what will you deliver, what value is to end-users, and its translations from vision.

Vision delineates where your organization wants to be and what it wants to achieve. To achieve that vision, you still have to translate it into actionable work. Strategy further defines this process by creating a roadmap of how you will get from vision to value.

Value streams take this process a step further by defining the daily work that contributes to strategy and vision, and therefore to value. Defining value on an organizational, team, and individual level helps you to focus work so you can measure its success.

As COO, your role directly involves linking company vision to work. This creates a firm connection between organizational vision, strategy, and daily activities. Work must connect to strategy so that outputs contribute towards specific and concrete goals. It's more important to work on the right things than to do things right. For example, if a technical team is spending too much time creating perfect code rather than working code, it could dramatically increase costs, moving the company out of alignment with a company goal.

Any and all work inside the organization must deliver concrete value. Connecting vision to work using value streams will help you define that process and how to measure it.

To get started, you must define what value is to your organization and to your customers.

What is value?

No matter how much work is being achieved, it's only useful when it's aligned with a business goal – some sort of deliverable that will bring value to the end-user. Unfortunately, it's not always simple to define value, as it can come in many forms such as a new feature, new report, a more stable product, better design, better customer support, etc.

In a traditional factory, you would normally purchase machinery and then put people to work building your product. SaaS is much more like building a product while simultaneously building a factory around that process.

You have to continue maintenance, development, and innovation around your framework, tooling, and infrastructure to continue to be competitive. This means that some people work to drive direct value for the customer (product or product feature) and others provide indirect value through infrastructure, forward growth, and maintenance. The relevant factor is that you need both to succeed.

Product and feature development are obviously the first place you see value. However, development isn't the only way you realize value. Some employees contribute by enabling other employees to do their jobs. An IT team managing information security is delivering indirect value by facilitating the creation of software and hardware.

So, you can define value as "deliverables that will grow your business". When your organization is developing software, value can be translated into product features that make your customers happy (or satisfy their needs). Other types of deliverables may be less concrete or more difficult to define in the short-term. Creating new processes, customer programs, training sessions, or a new customer support service may not directly contribute to the customer's immediate happiness, but they will create an environment which enables you to deliver better value.

Defining "what is value" inside your organization allows you to link work to actual deliverables. This, in turn, enables you to monitor the value and success of work.

As COO, Scrum master, etc., it's important to monitor how much time is invested into enabling growth and value and how much is invested into direct output. Why? Enablers, or work that enables other work, are valuable, but only to a certain extent. If too many of your efforts are focused on keeping the lights on, you won't be able to provide new value to the customer.

How do you define what is too much? I think it's too much if you are not able to deliver to the expectations of your stakeholders, as defined by the 7 Delivery Qualities.

It's relevant to understand what kind of work people are performing, so you can measure and monitor it. This will eventually help you to understand if you are doing too much maintenance or work that is unrelated to direct value.

> **Strategy Tip:** Define what value looks like for your organization, so that you can use it to monitor throughput.

The Value Stream

The value stream is the process, or sequence of steps, that creates the bridge from vision to value.

Value streams are a concept originating in Lean management, which directly map areas of work to added value. A value stream is a channel which you define to deliver value to the customer - starting with an idea or assumption and going through a pipeline until it is delivered and reaches the end-user. You can think of it as sort of an assembly line, where you recognize the demand for a product or feature and bring work together to generate value for the consumer, with product development, customer service, organizational development, etc., each contributing to the result.

In tech companies, value streams mostly relate to the product development pipeline. Your organization has an idea for a new feature and creates a value stream leading to that feature's delivery to the customer. However, this isn't always the case. Value streams can be constructed around anything that delivers value. So, an internal organizational project, such as a development track, a new finance approval process, or other projects for customers like a new support program, or e-learning material can have value streams.

A value stream defines the most important steps in the pipeline, which means that it will change depending on the pipeline. In the scope of a tech company, the value stream maps someone sending a feature request, it's assessment, planning, estimation, development, quality assessment, and deployment.

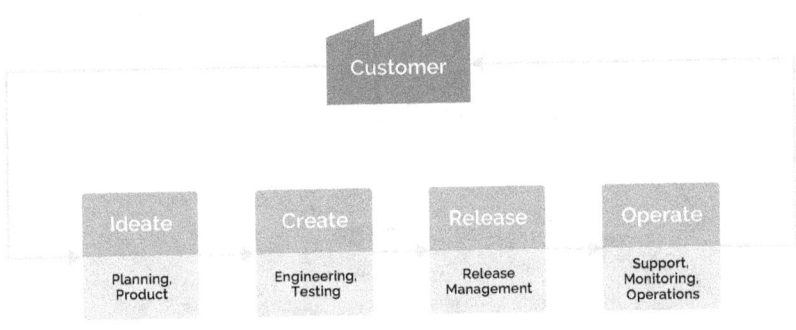

Fig.7. Example of a Product Development Value Stream

It's simple for tech companies to identify value streams when mapping product development. However, many begin to struggle when mapping work that is not directly related to software features. Producing knowledge base articles, a customer program, or a new internal service will generate value and can have value streams. But, even identifying value streams for product development can be complex. What happens when you have different products? Should you have separate value streams? Or should you split teams within the same value stream?

Of course, there isn't a simple answer. Value stream mapping is entirely related to how you prefer to organize your operations. For me, the most important factors are the end user, how you deliver value to customers or to employees, the nature of deliverables, the purpose of teams, and work processes.

At Nmbrs, we began with 4 value streams. We used Organizational, Market, Nmbrs Payroll Product, and Nmbrs HR Product value streams.

The Organizational value stream defined the processes of work to develop the company, in the scope of HR, Operations, and Finance. Here, output is value for employees, such as introducing training programs, onboarding tracks, new roles in the organization, or an office expansion. The Market value stream orients to prospects and relates to new business development, with most activities falling under the scope of Marketing and Sales. Deliverables in this value stream are items such as a new website, marketing campaigns, blog posts, or live events. The Product value stream was split into the two main applications we are developing.

Each value stream has its own teams and scrum-of-scrums. We also noticed that it would be more beneficial to the Development value stream to have more coordination and knowledge-sharing between tech teams, so we merged the two. At another point, we realized that it would be beneficial to organize our teams along the full customer journey, so we made the decision to merge the Market and Product Value streams, considerably improving the collaboration between sales and tech. Eventually, we ended up with just two distinct value streams: Organizational – to deliver a better organization to employees – and Product – to deliver a quality product and services to prospects and customers.

Value stream mapping is the practice of designing the steps in a value stream around business processes. Mapping the steps of your value stream and measuring added value and waste time will give you a better idea of how long it takes to deliver value once demand begins. It also enables you to identify waste time and bottlenecks in the process so you can work to improve.

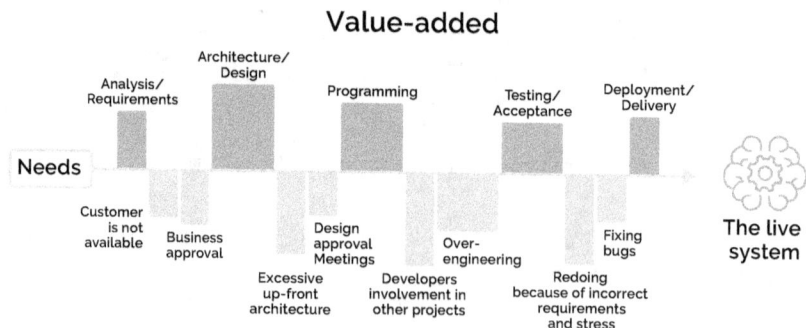

Fig.8. Waste time in the value stream

Agile organizations typically pile work items onto a backlog. Here, they wait to be prioritized or wait for capacity to pick them up. This can result in "waste time". What is waste time? Waste time exists when no one is actively working on product development. Instead, they are waiting for stakeholders, approvals, for other teams (dependencies) re-work due to poor implementations, etc. Backlog items contribute to this because they may influence other aspects of development or another team's ability to move forward.

Your value stream should contain processes to monitor where time could be wasted, so you can optimize delivery.

> **Strategy Tip:** Map your value stream and measure waste time and bottlenecks to optimize your processes.

Measuring Value

Measuring value allows you to improve operations with prioritization. If you can prioritize work based on how it adds value, you can better direct operations. Creating value streams that align with long-term vision and organizational goals and measuring value based on "end" customer satisfaction will help your organization achieve its goals through work.

Measuring value involves identifying KPIs that define and measure progress. These KPIs should be aligned with the business plan, organizational strategy, and company vision. This ensures that what you are measuring is actively contributing to the future success of the organization.

Your process of measuring value should align with how you prioritize indirect and direct value. Product development often makes it difficult to measure throughput, or the maximum rate of production. On one hand, it's easy to quantify specific metrics such as so many lines of code, number of solved tickets, or the number of updates delivered. On the other hand, none of these performance indicators necessarily directly contributes to the value stream. All the code could be related to maintenance or internal factoring instead of new features. Even lines of code related to new features could result in features which don't contribute to the value stream because users don't want them or don't find them relevant.

For me, it is most interesting to have insight into metrics relating to rate of delivery for work related to organizational goals. These could include flagship features or those with direct impact for end-users, maintenance, or internal features.

Implementing processes that measure value means getting teams to understand what does and does not contribute towards high-level goals. In my company, we chose to directly involve teams, asking them to mark epics as either "new feature" or "maintenance". New feature directly impact value for the user, in the form of a new feature or improvements to existing functionality. Maintenance relates to internal work such as code refactoring, improving non-functional requirements like performance or security, and bug-fixing. This divide means that maintenance also delivers value to the end-user, but in an indirect fashion.

Our Scrum masters maintain dashboards that monitor team's delivery of maintenance vs. new features. This allows us to reduce total maintenance time for teams, so they can focus on direct value. This gave us a better grasp of how and where value was being delivered.

> **Strategy Tip:** Create KPIs to track how much work delivers direct value vs. indirect value.

Maintenance Prioritization

It's easy to get caught up in the day-to-day work of maintenance, attending meetings, and fixing problems as they arise. While this work is often necessary and valuable, it shouldn't normally be your priority.

Simply working to "keep the lights on" will not drive the long-term value to grow your company. Bug fixes, support incidents, and other short-term solutions are important, but short-term satisfaction will eventually become long-term frustration.

If you're spending all your time patching issues, you're not devoting time and energy to what's truly important – developing long-term solutions to problems, creating new features, and a better user experience. Or, you're focusing on small problems first and big problems, like creating a better user experience, second.

A good analogy here is a popular story, "The Rock, Pebbles, and Sand"

"A philosophy professor stood before his class with a large jar. He filled it to the top with large rocks and asked his students if the jar was full.

"Yes", they said, the jar was indeed full.

The professor then added small pebbles to the jar, shaking it to disperse the pebbles between the larger rocks.

"Is the jar full now?

The students agreed the jar was full.

Then, the professor poured sand into the jar. The students agreed the jar was completely full.

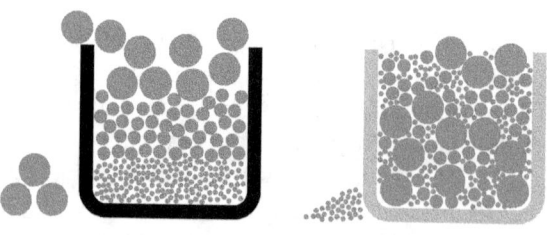

Fig.9. Jar metaphor (Prioritization)

In this analogy, the rocks represent big and meaningful things, like future product development, customer satisfaction, and achieving long-term goals. The small pebbles represent less important but still meaningful tasks, while the sand represents the small day-to-day work of maintenance. If you fill the jar, which represents resources, with sand first, there won't be any room for the larger, more important items.

Prioritizing resources to fit in the most important items first (development and customer satisfaction) and fitting less important tasks around that will help your organization to support your organization's growth.

Time allocation is a standardized practice for helping teams manage and prioritize how and where they spend time. For example, if your team dedicates 80% of time to developing new features and 20% to maintenance, the team has a guideline of how much of their day-to-day efforts should be spent on each. Ratios used by teams should depend on factors such as quality, alignment with higher-level goals, and season. Introducing ratios will help Product Owners prioritize their team's time while improving estimates of how much maintenance their team is actually doing.

Maintenance is an important part of continuing to deliver quality, but it does not contribute as much to delivering value to the customer. Therefore, your focus should remain on developing new features or improving the product to reduce required maintenance.

Strategy Tip: Begin with and prioritize significant projects, don't get lost in maintenance.

Stakeholders

Stakeholders are persons with an interest, concern, or who are impacted by something, which can be your organization, the value you drive, or specific business processes. Stakeholders can be internal – such as employees, specific teams, or your management team – or external – such as investors, customers, end-users, or partners.

Stakeholders should always be involved in value streams so they can collaborate, work to guide the value stream towards its end-goal, and drive value for the end-user. Engaging stakeholders as a continuous part of your process ensures they can contribute to the process, keep teams on track, and work towards the end-goal. This helps to eliminate waste while ensuring efficiency because you know the value stream is delivering what the stakeholders need.

Selecting stakeholders for important business drivers ensures you have an interested party keeping each element of the business on track. However, it's important that stakeholders understand their purpose, what they represent in the value stream, and what they are working for, so they can achieve value for the organization.

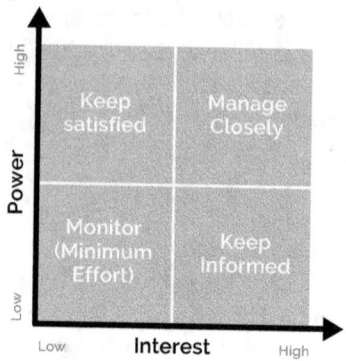

Fig.10. Stakeholders Matrix

It's important to keep stakeholders engaged and involved with individual work. Teams are responsible for delivering value through product development and customer support, but stakeholders should always be part of the decision-making process. Stakeholders should be involved in work prioritization, choosing which projects are picked up by which teams, and so on – providing they have the working knowledge of the organization to make good decisions.

At Nmbrs, we use a structure with no managers. We involved stakeholders by defining each Agile squad and chapter as a potential internal stakeholder. Each squad has its own purpose, so it's clear what the squad is aiming for. However, because the scope of the squad may not be broad enough to deliver the full purpose or objective, work completed by one squad might need to be taken to the Product Owner from another squad. Here, the squad takes a stakeholder role, delivering their own work or allocating work to another squad.

We also implemented Chapters, groups of specialists who belong to squads, as internal stakeholders. For example, all developers are grouped into the Development Chapter and the QA engineers are in the QA chapter. These chapters are responsible for quality guidelines and strategies for that work. So, when the QA chapter rolls out a QA automation strategy to all squads, the QA chapter becomes the stakeholder of that work, so they can oversee development, implementation, and changes.

Stakeholders should be consistently informed of work status, delays, delivered features, and progress so that they have a clear view of the value stream and work being completed.

Strategy Tip: Select a stakeholder for each important business driver and involve them in value streams.

Connecting work to vision

Many organizations develop long-term vision and strategy at a high-level and only share it with management. I strongly believe this is a mistake. Connecting daily work and high-level projects to purpose and vision ensures that individuals understand not just what they are doing, but most importantly, why they are doing it.

When people understand the vision behind their work, they are able to work in an Agile way, creating their own ideas and solutions framed within the bigger picture of the company vision. Giving your teams this kind of freedom will increase motivation while helping you scale teams. Most importantly, you will notice individuals taking ownership of problems and taking responsibility for finding solutions, because they understand how their work contributes to end-goals.

One way to connect daily work to organizational vision is to structure work detail into layers, such as Epics, Initiatives, and Goals. Some Lean frameworks break this down into Themes, Initiatives, Epics, and Stories.

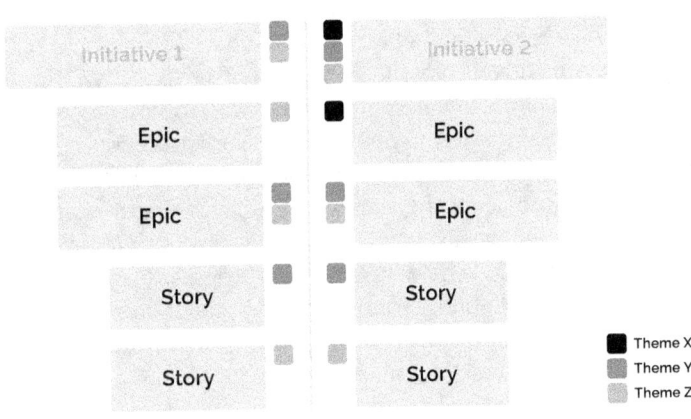

Fig.11. Aligning High-level initiatives to daily work

Here, an initiative is a high-level collection of epics driving towards a common organizational goal (such as producing a new product or extensive feature set). An epic is a large body of work such as an entire feature. An epic can be broken into smaller bodies of work called stories, which would be sub-parts of a feature. Stories are groups of work which can be assigned out at an individual or team level.

Organizing work in this way will improve task-based motivation because individuals can take ownership and steer work towards a defined organizational goal. If everyone knows what they are working on and how it connects to larger goals, they will have much more control, input, and idea of what they are trying to achieve.

Connecting daily work to organizational vision enables you to see if you are moving from vision to value. When you can directly break vision into smaller and smaller work pieces, you know exactly how you are achieving vision.

You should be able to trace any work to high-level business goals. Even if you're looking at the smallest unit of work, such as a ticket for a task or user-story, it should relate to a high-level goal. When you cannot, that work should be considered maintenance or work done out of the scope of your higher-level focus or objective.

> **Strategy Tip:** Make sure there is a connection between daily work and vision.

Standardizing Work Processes

Work structure becomes more complex as you scale and grow. Processes that once belonged to a single team will eventually be owned by multiple teams. It's normal for every team to work using their own process in startups, but this becomes a problem as you scale. If every team has their own work approach, it's difficult to monitor work output. You won't know how and where each team is contributing to work. Maintaining work at a standard of quality, to similar production, and within budget across the organization is a priority. Doing so as you add on more teams necessitates standardizing work processes.

One interesting practice when standardizing work processes is to abstract all work into something that can be seen and handled like a product. This enables you to create a common concept which can be allocated to a value stream and to a team. This will, in turn, allow you to develop a backlog for improvements, identify end-users and stakeholders, and track progress based on measurable output. With the ability to budget based on the product, to create work standards for the product, and to create a single product owner to have a final say and responsibility for all product output, this tactic gives you much more control over the work of multiple teams.

This very broad product concept can be applied to virtually anything within your organization, from software products and modules to processes, services, internal assets, and much more.

It's a good practice to identify all products within your organization, so you can map value streams, what is attended, and by whom. Why? Everyone naturally works in different ways and you likely have tech teams working on applications and code while marketing works with blogs and websites, finance with spreadsheets, and so on. It is very useful to abstract everything to "working on a product" so you can unify work items. You can then treat all output as though it directly contributes to a specific output (because it does). For example, you can create a website as a product, allowing marketing and tech to work on it together, each with their own defined goals and output. You can then link it to standardized work processes, rather than allowing everyone to complete work in their own way.

When you deliver a new product or service to your customers, it can likely be translated into one or more internal products which you can allocate to one or more teams. Complex services will naturally span several products on several teams but allocating such services to one team will simplify the process.

Using internal products offers the most advantages in terms of budgeting. Why? It allows you to allocate budget towards the product as a whole. This reduces the need to give budget to teams. In short, it gives you the resources to track how much you invest in a specific product (website, blog, model app), even when the product changes hands and another team takes over or when numerous teams are working on the product.

Creating internal products also allows you to better define product ownership. The larger your organization, the more frequently change happens and the more common it is for some products to be unattended. When any product is unattended, no one is responsible for ensuring its success. This is something you should resolve whenever you find an unattended product.

Finally, it's important to keep internal products granular and at a similar size. You don't want some internal products to require a few days of work and others to require months or even a year of work. Consider factors such as workload, maintenance time, etc. to better judge size and define what should be an internal product inside your organization.

> **Strategy Tip:** Identify and map all internal products within the organization.

Work Visibility

Everyone inside of your organization is working and busy. Whether behind the computer, developing software, calling customers, attending meetings, etc. people are doing and producing something. But, is that work contributing to organizational goals? How can you measure if work is moving in the right direction, according to company vision, goals, and strategies, and moving at the right pace to achieve those goals?

Having busy people inside your organization doesn't necessarily mean your company is being productive. People who work around the clock to compensate for chaotic environments, or who have to redo work because of miscommunication or misalignment are not being productive, even if it seems like they are. You have to assess the direction of work, progress, and productivity.

Designing processes to measure value won't have any impact if you can't see work in the first place. If you want to measure value, you'll have to make work visible. What is visible work? It's work that has been clearly identified, assigned to a team or person, given a clear status and progress updates, and connected to a higher goal or strategy whenever possible.

Making work visible can be very challenging in organizations where teams do different types of work, such as development versus sales. This is especially important where people register work in different ways or when some don't register it at all. Zooming in to a macro level will keep bringing up more and more differences between how people work. However, if you zoom out, work from a high and abstract level, and go from there, it is possible to make work visible across the organization. Patterns will appear and you can use them to group and track work. If you zoom out far enough, you will likely see that many people are working on "projects" or doing "maintenance".

If you can define what a "project" is inside your organization, you can better define what those people are working on. For example, how long is it, how long does it take, how can you describe the work, can you register it on one platform? Once you achieve this, something amazing will happen – you'll see what your organization is working on!

Work visibility will help management drive the organization, but it also helps teams and stakeholders to understand what's happening, who is working on what, and which obstacles may prevent progress. Activities such as task prioritization and assigning work will become easier as well because you'll have a clear picture of what needs to happen and what is happening already.

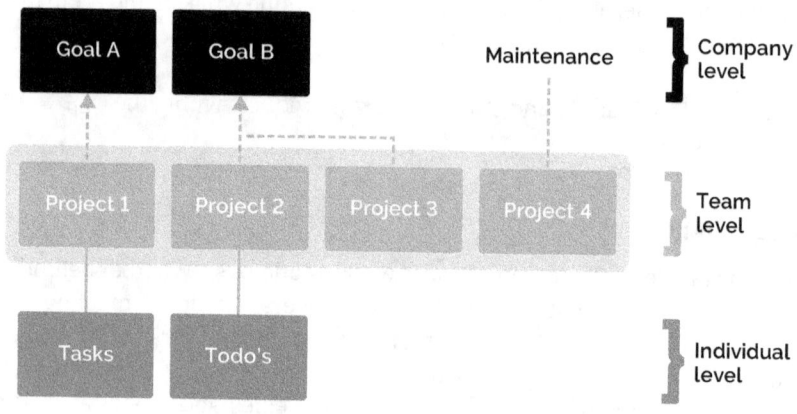

Fig.12. Work visibility layers

It's important to identify organizational-level goals or common factors for everyone, which you can then break down into team projects and individual tasks. This allows you to create broader-scope goals visible to everyone and then define details where it's relevant.

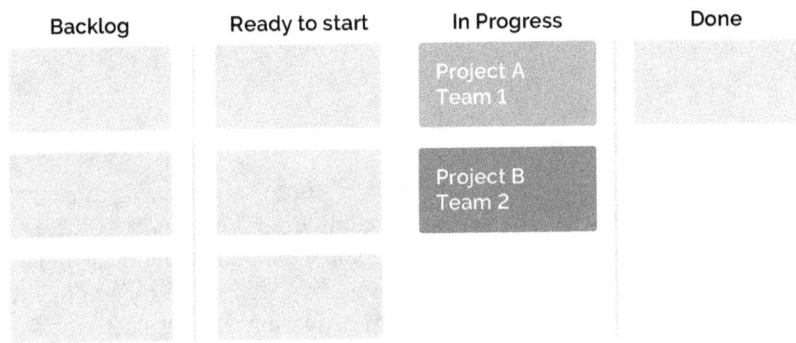

Fig.13. Work visibility progress

This type of overview will enable you to track work in progress, while creating a clear overview of backlogs and work finished for the entire organization. So, you could track each team's contribution to projects or epics, track individual contribution, and see how everyone is moving the organization to a common goal. Identifying units of work such as "projects" or "epics" will also give teams a good sense of deliverables and progress, which can help with motivation as well.

One of the most important considerations for creating a visibility framework is ensuring that people actually use it. It's quite common that operations will create a structure and integrate their visibility framework into tooling, only to find that no one is using it. You can have a perfect board and not have any data, because people just aren't registering anything.

The process of making work visible is an extra step in the work process. Your teams have to be aware that it is a necessary step. At Nmbrs, our Scrum masters have to put a lot of time and attention into ensuring individuals register when they start, stop, and complete work.

Strategy Tip: Make sure all work is visible in a standardized way.

Once you have a work visibility framework in place, you can use it to add further value. For example, you can flag impediments such as bottlenecks and dependencies, unclear responsibilities, unclear strategy, etc.

Why? Most people will simply stop working when faced with small hurdles such as waiting for approval, waiting for other teams, unclear strategies, or when too busy. If your work visibility framework includes a system to flag when work progress is not optimal, people can help.

This can be achieved in numerous ways. For example, you could have a chat channel for impediments or standups. It's important not to rely on managers or seniors to identify and handle impediments because they often are the bottlenecks. Ensure everyone can flag issues. These alerts should be visible, transparent, and registered so you can identify and monitor items slowing your organization.

Strategy Tip: Create a system for people to flag impediments visible to the entire organization.

Prioritizing Core Processes

Any organization is built around a series of core processes which are mission-critical to the success of the organization. In tech companies, these processes typically revolve around product updates and deployments, which are crucial to delivering products to customers. Over time, you may notice that these processes actually prevent you from delivering products with the same velocity or efficiency – even when nothing about them has changed.

There are often key points when you realize that delivery has slowed, efficiency has dropped, and things aren't running as smoothly as they used to. In my company, this type of problem typically relates to core processes which are no longer working properly, or which are no longer suited for the number of users or the technology.

This happens even when the process seems to be okay, especially with long-standing processes. Finding issues can also be slow. Users run on autopilot while completing the process they are overly familiar with and will be slow to recognize errors and issues.

Eventually, problems of this nature will slow and may even stop production. Sometimes it takes weeks or months, but it will happen. Most crucially, process-related problems are silent issues, as teams become accustomed to them and will not likely report them as a critical problem. Everyone is busy running the process, so no one is focused on improving it, and no one notices when it's not running smoothly.

Even if your technical value stream is running perfectly, developers writing code will have no way to deliver the product if the deployment process is not working. Processes such as deployment are critical to organizational success, but they are often seen as secondary features or functions – rather than as essential to the delivery of the product. Problems with these core processes can bring everything to a halt when something goes wrong.

When the core process of delivering product updates fails to function optimally, it creates a bottleneck for the entire organization. With no way to deliver or to deliver well, you cannot deliver value, which means there is no point in continuing development. Fixing this core process issue is key to running the entire value stream.

Any instance where a core process is suffering is a high priority. It will interfere with every other level of the organization. If your test environment is not reliable, your continuous integration system is not working, or your team no longer trusts your testing framework, it will hurt productivity and efficiency. Core processes must run like a well-oiled machine and that means staying on top of how they function, ensuring that they always run smoothly, and that they are always up to date.

Here, many organizations will delay fixes to core processes, simply because they temporarily slow down or delay operations. However, short-term delays are significantly more affordable than the costs incurred by an inefficient process, which could impact work for years.

It's important to develop a framework that enables you to constantly measure and monitor the health of core processes. This will enable you to recognize when issues occur so that you can act quickly and solve them before they become problematic. Symptoms of core process issues often include teams taking longer for the same tasks and processes, unexplained decreases in quality, and increases in escalations. Map your KPIs to symptoms of core process failure, like deployment times, test environment downtime or code commits conflicts, and you'll have an easy way to detect when you're slowing down.

Strategy Tip: Make sure your core processes are always running efficiently and prioritize fixing them.

Solving problems while the train is running

Working on software development, especially for SaaS products, can feel like being on a train that never stops. You have to update every day/week/month to continue delivering value and to keep the product running optimally. Anything that you do inside the company – from maintenance to updates to problem-solving – has to be done while that figurative train is running. This necessitates creating a process which enables you to push updates while running core processes, so that when you develop and launch big features, this maintenance doesn't interrupt your core software or its performance.

Why? Your application always has to be available to support existing customers. You have to provide services like customer support and problem solving. You need to process invoicing. You have to continue to push software updates and fixes to keep your product running optimally. Simply maintaining your software requires constant work and stopping any of those activities would be stopping your business.

At the same time, you have to work to develop new features, new iterations of your software, and big product updates. These improvements to services, activities, and products are necessary if you want to keep your product on the market.

Let's consider an example. You need to define a better way of organizing the visibility of your teams' work but to do so, you have to reformulate the epics and tooling they are working with and introduce a new product roadmap strategy. If you ask teams to stop working while you integrate changes, your delivery value stream will grind to a halt. How can you integrate changes with as little disruption as possible?

In most cases, integration should depend on the case at hand. You can work small improvements into the production schedule, with minimum disruption.

Larger updates require a more structured approach. Here, I like to use a branch approach, creating a parallel flow to develop and integrate solutions and features, which eventually integrate with the still-running process or activity. With multiple touchpoints to validate and test new features before integration, this approach enables you to integrate everything smoothly and with as little disruption as possible.

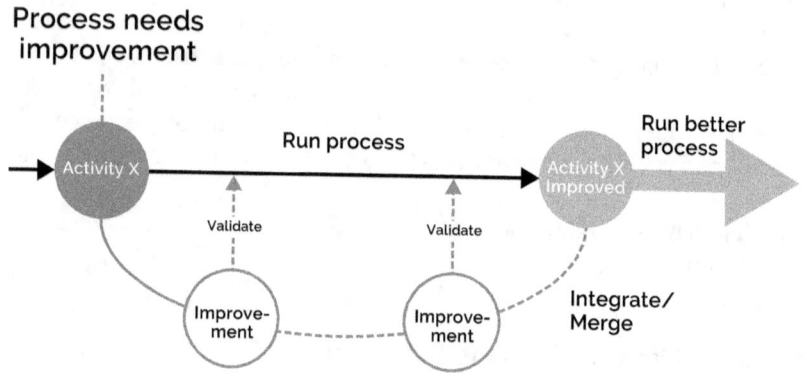

Fig.14. Introducing changes on a running train

Creating a parallel flow means developing an implementation that you can validate against the real process. This enables you to create iterations which are tested and validated as part of the development process, so you can ensure your final product is ready. Incorporating changes into the product is always the most difficult part of the process, but having a previously validated solution will minimize the impact of integration.

This approach works for both product development and pushing fixes. In fact, it's inspired by code branching techniques, which are commonly used by development teams when changing substantial portions of source code. For example, if you need to rebuild a feature your customers are currently using, a parallel flow enables you to develop and validate the new code without disrupting customer service.

At a high level, branching your development is simply the process of creating a side activity where you work on the improvement or update, validate it at set times, and integrate into the running process or activity when it's complete. This process is extremely valuable in situations where you have to integrate a new solution for processes or activities but cannot stop the process or program to integrate it.

Strategy Tip: Develop processes to ensure you can make improvements or deliver fixes while the processes are running.

Focus on root fixes

Solving organizational problems – whether small ones relating to single development issues or larger process and infrastructure issues – necessitates that you review the problem to determine what's actually wrong. While it's easy to go about solving problems, you aren't actually achieving anything unless you're solving the root problem.

Identifying root problems often means approaching any problem as being a symptom of a bigger cause. The "5 Whys" is a useful tool to help you identify the root cause or causes behind a problem. First introduced by Sakichi Toyoda in the Toyota Motor Company, the "5 Whys" became critical in problem-solving training and eventually in the Toyota Production System, forming the basis of Toyota's scientific approach. The method has since spread to other companies including Semco (which practices the 3 Whys), and is now integrated into Kaizen, Lean, and Six Sigma.

Here, you continue to ask "Why" until you identify a root cause, or a problem creating the other issues. This may require more or fewer than 5 "Why's" but is a valuable tool for approaching any problem as symptomatic of a greater issue.

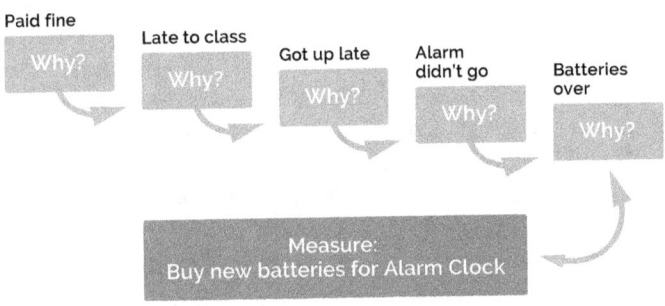

Fig.15. Identifying the root cause with the 'Why's' process

If you receive a lot of customer support tickets, it's easy to solve those tickets and to feel accomplished. You're making customers happy solving the problems. But, because the instance that the customer contacted you about is still an issue within your system, more customer support tickets will keep piling up. This is especially important in the SaaS category, because solving symptoms will eventually put you in an endless circle of patching issues. You need to determine where the root issue is so that you can permanently solve the problem instead of the symptom.

Let's say that one of your team members is not performing. You could use the following example of the 5 Whys to determine the root cause of the problem.

Symptom	Direct solution	Why-reasoning
A support consultant is unhappy and tired.	Send him on vacation	Why was he tired?
He's unable to handle all his incoming tickets	Hire another support consultant	Why? Are there too many tickets or is he unable to solve them?
It takes too much time to solve incoming tickets	Hire another support consultant	Why? Does he lack knowledge? Or is the content very complex?
It takes too much time to investigate the user's problem in the product	Create desktop-sharing to help support agents find the problem with the customer	Why is it difficult to find problems in the product?
Users make a lot of mistakes and it takes time to figure out what went wrong	Improve user documentation in a Knowledge Base	Why do users make so many mistakes?
The product is very complex and error-prone	Improve the product	Why is the product error-prone or difficult to use?
There are too few User Experience experts in the team	Hire UX designers!	(problem solved) (if the problem is not solved, keep going)

In this example, a stressed and unhappy support consultant uncovered a product weakness, which, in this case, is usability. This exercise could go on and on, possibly identifying further root causes such as budget limitations or using the wrong hiring strategy. The deeper you ask "Why", the more you will discover if your strategies are aligned with your goals and if or where you have bottlenecks in your operation, services, or products.

Most importantly, if you were to apply the direct solution before discovering the root cause of the problem, you would just be fixing the symptoms. Sending a support agent on vacation will only result in him returning to a stressed point after the vacation. You haven't solved anything and the same problem will continue to recur. Or, if you were to continue hiring support consultants at step 2 or 3, commercial success would dramatically increase your support costs because the flaw is in product design, and you would have to keep hiring to meet growing needs. Solving the root problem, in this case UX design, is a much better and more cost-effective, long-term solution.

One of my favorite analogies here is that of a water leak. If you have a bucket full of water and it keeps leaking, you can keep the bucket full if you keep pouring water in. But, unless you take the time to find the hole and patch it, you'll be stuck in an endless round of refilling the bucket.

This also applies to escalations, where critical process failures such as data leaks, security incidents, application failure, or similar problems occur. How you react to these problems will greatly determine your end-results. If you immediately solve problems caused by an application failure (the program is offline), you won't fix the problem and it will likely go offline again very soon.

Here, it's crucial that you have processes in place to clearly define what should happen when a critical escalation (often called a code red situation) happens. You should have defined types of incidents, defined response teams to handle those situations, and a clear way to alert the organization when things go wrong. This process must include ways to prevent more damage from happening by recognizing the root cause and working to fix that as the initial problem is solved.

These types of situations are always disruptive for teams and goals. They will impact deadlines and customer expectations. Everyone will have to drop planned work to focus on an emergency. Your escalation protocols should also include built-in processes to communicate with stakeholders and customers when necessary.

Your strategy should be to create an approach that works to identify the root cause of escalations, fix it with high priority, and communicate all information as it's available to stakeholders. The process should both communicate urgency and the need to solve the root of the problem by only accepting first-time incidents. Any SaaS company will face recurrent issues, so it's crucial that you be able to identify and fix the root cause while improving monitoring to prevent future occurrences as part of a code-red solution. It's also important to schedule a retrospect session after any code red situation, so all teams involved can work together to create an action plan for preventing the problem from happening again.

Solving problems, code red or standard, means creating processes that approach the problem from a root cause. You also need communication, clear protocols defining who to contact and who is responsible, and an approach designed to solve the root cause, rather than covering symptoms.

Strategy Tip: Identify the root cause of escalations and prioritize fixing them over normal work and before solving symptoms.

The power of routines

Routine helps people approach tasks more easily and more comfortably, because they are accustomed to them and may even perform on autopilot. It's important not to underestimate the power of routines because they can help you in building good processes and culture. For example, consider your morning routine (wake up, bathroom, shower, get dressed, eat breakfast, prepare your work things, go to work). You handle everything on autopilot, likely with little or no thought, because it's a routine activity.

In a work environment, routines can help people to follow certain activities that are important and that don't need to be discussed each time they're completed. These activities will eventually become part of your culture.

While routines can be extremely valuable, they are difficult to start. Routines are created by habit, so you or someone else in your organization will have to push desired activities long enough for people to perform them naturally, even on their busiest days.

Consider daily team stand-ups. Teams that don't do them will struggle with organizing and even showing up. No one will know the time, where to go, who will join, or what the agenda is. It can be like organizing an event! But, when it's done regularly, routine kicks in. People automatically show up at the same time, in the same place, with the same people and agenda as before. It doesn't require planning or thought, it's just done because it's a routine. Once established, a routine like this one will help teams to focus on the content of the stand up and not the agenda or planning.

The best way to develop routines is through consistency. Push for the same activities at the same time, define a heartbeat, and your company culture will grow around routines.

> **Strategy Tip:** Set and consistently push for routines for important work activities.

Work cycles are the heartbeat of your organization, they define the movement of your company and the processes that bring it to life. Heartbeats and work cycles can define everything from high-level cycles such as company seasons which may occur yearly or semesterly, to quarterly cycles defining goals, to low-level team sprints.

Defining these cycles at different organizational levels will help teams organize and define the scope of work, define milestones, and deliverables. This will support establishing processes which enable work to be completed in a way that delivers value.

In most cases, you should begin by defining short-term, long-term, and mid-term work cycles or layers. A common tactic is to define year-cycles for long-term, quarter-cycles for mid-term, and two-week sprints for short-term. You can then nest shorter layers inside of longer ones so that work cycles are defined and mapped for the short to long-term, and you can easily see what is contributing to each larger goal.

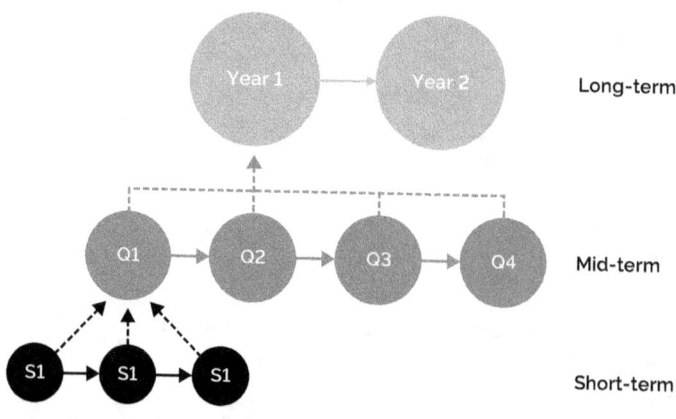

Fig.16. Work cycles and heartbeats

Most work will eventually be defined as a sprint, framed into a mid-term milestone, which is then connected to a long-term theme. You create short-term goals with a focus on mid-term goals and aligned with long-term company vision and strategy.

> **Strategy Tip:** Define cycles for all organizational levels.

Once you've defined work cycles, you can begin to define the heartbeat which ensures that cycles continue to flow. Here, you should hold a kick-off session with relevant teams and stakeholders, defining the purpose of the cycle and the expected results. You can repeat this at all levels, with relevant people at each stage.

Long-term cycles should be discussed with the management team and shared with the organization as a whole. Here, organizational focus and goals and their relevant business impact are defined and made visible to the entire company.

Quarterly kickoffs are useful for groups of teams collaborating across projects, applications, or markets. Here, you should involve the relevant teams and their managers, but not the entire organization.

Short-term cycles can be planned on a team level, where the team comes together to define the sprint, its tasks, and stories. This is a common Agile practice, and one that I strongly encourage.

While planning and creating a heartbeat for a work cycle is important, it's also a good idea to follow up. When cycles are completed, host a session with the original kick-off members to discuss work completed within the cycle, to identify learning points, and to discuss what could be improved for the next cycle.

Cycles come and go in cadence, along with kick-off and retrospective sessions. It's always a good idea to plan the next session or step during the current session, so that everyone already has a specific planned date for the next step. This will help with communication because everyone is aware of the cadence of the heartbeat, or when and where they have to follow up.

> **Strategy Tip:** Always plan the next step to keep the heartbeat of your organization steady.

Work dependencies

Work dependencies are one of the biggest productivity bottlenecks inside any organization. Here, someone is unable to continue work because they are waiting on someone else, either to finish work, make a decision, give permission, approve, or give feedback on something.

Work bottlenecks are very often invisible because most people will simply start something else while waiting. Unfortunately, people and teams switching focus to work on new things aren't working on what is likely their highest priority

– the work they're waiting on. The end-result is often that everyone is working but deliverables are not being met.

How do bottlenecks like this happen? Many organizations, especially larger ones, require certain activities to be approved by a manager at each stage. This often ties into security, to ensure segregation of duties and principles, or to prevent fraud and errors. These types of structures are often required by auditors and compliance officers to ensure the organization remains compliant with quality and BPM standards.

However, integrating too many dependencies may considerably slow productivity by creating single points of failure. For example, when a certain key-person is not available to give approval for a critical process, the process is delayed.

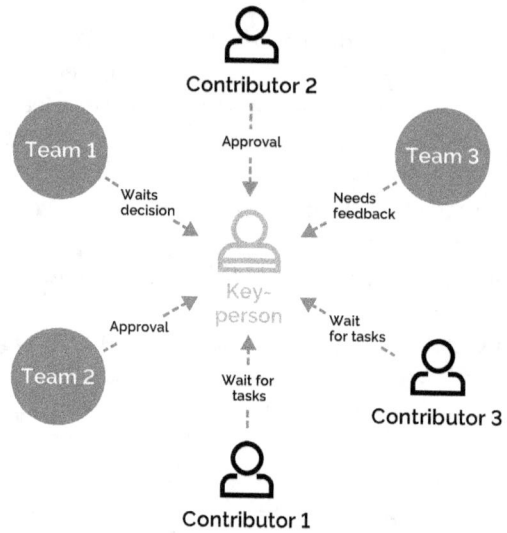

Fig.17. Work dependencies around a key-person

Dependencies also relate to work output and responsibilities. For example, if one software development team is responsible for the backend of the application and another for the front end, you need both to work together at approximately the same pace. If you want to deliver a full new feature, let's say a new report that displays a list of users, you will have to coordinate both teams.

Here, the front-end team would be responsible for creating the new user interface, user interaction, input validation, etc. of the features which will display the data provided by the backend team. This creates a dependency for both teams, because the front-end team can only deliver and test their work once the backend team finishes theirs.

This situation becomes more complex when teams waiting on work from other teams start new work. This could create a new dependency with a third team but may also slow development and deliverables for the original project, because the team will finish what they're working on before going back to the original project. At the same time, this could create a new bottleneck, where the back-end team is waiting on the front-end team or develops a sudden critical incident and shifts attention to focus on a new task. Bottlenecks can eventually create a very long dependency chain which will be difficult to untangle.

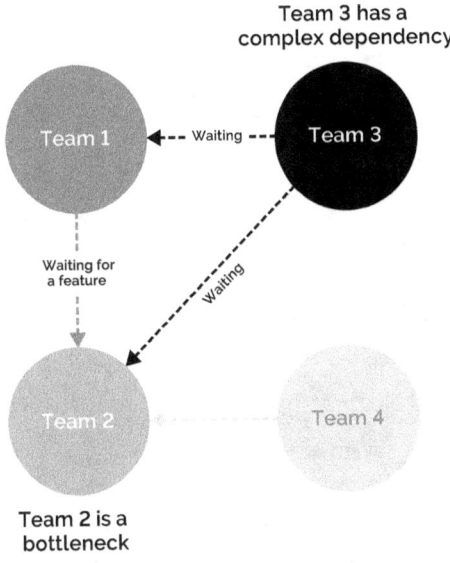

Fig.18. Work dependencies on teams

A third type of dependency occurs when teams begin tasks with a certain scope which then changes. For example, your team starts developing the same report feature mentioned above and, while developing it, they discover that a new UI component should be developed to minimize future maintenance time. In this case, they could either make the scope of their work bigger to include the component or begin going out of scope.

Keeping this type of project in scope would require creating a new work item. This could then be allocated to their or another team's backlog, to be completed after the current project is finished. Then, if new problems arise – such as the team finding that graphic and style assets are not centralized – while developing the new UI component in our example, the team could create a new project to enable the new feature without derailing or bottlenecking their priority goal.

Instances where new work arises from existing work should always be identified and flagged for stakeholders and managers. What should be done? It often depends on the situation and whether the new work greatly affects what's being completed. Depending on the situation, you can enlarge the scope of the original task to improve the result or stick with the original story and create a backlog for later.

Having too many of these dependencies will increase the complexity of project management, so it's always a good idea to minimize them wherever possible. Instead, create backlogs and assign work elsewhere whenever possible to keep projects on scope.

Strategy Tip: Minimize work dependencies involving processes, roles and responsibilities to maintain productivity and meet deliverables.

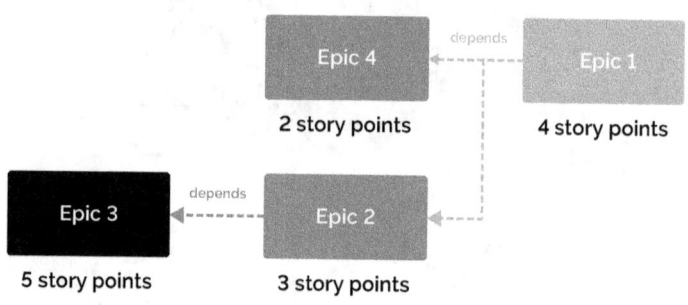

Fig.19. Work dependencies on projects

It's also important to keep in mind that backlogs can affect other work. In the case of the new UI feature, it would affect the original deliverable, creating a dependency. Backlogs with dependencies should always be visible and properly shared with stakeholders, so involved parties can see how work is progressing and where it might be slowed. Deciding on certain work items may slow work down more than initially intended, which can affect budgets and planning, as well as the final deliverable.

During project kick-offs, try to identify any dependencies between teams, components, technology, etc. so you can better predict timelines and delivery.

Strategy Tip: Work to identify dependencies before starting a project and make them visible.

At Nmbrs, one of our primary mitigation factors for team dependencies is to ensure squads are comfortable working on each other's source code.

Introducing this process into your squads can be met with resistance. Squads won't always have the domain knowledge they need to successfully update the entire code. They feel like it's not theirs, so they don't dare touch it. Squads also need a lot of support from QA engineers when working on another squad's code to validate if those changes have an impact. Without someone there to walk the squad through changes, everything comes to a halt. However, when you're building large features with work split between several squads, you will always have these dependencies.

The solution is to rely on QA automation. This ensures that squads can always validate changes to another squad's code, without having to wait until that squad is ready to help.

Let's say Squad A is making a change to their code. They have to update Squad B's source-code to ensure compatibility. They do so. Squad B has implemented QA automation, so Squad A introduces the changes and runs automated testing on the server without involving anyone from Squad B. There is no dependency. Squad B can always step in later to make small changes or fixes, without forcing Squad A to wait the entire time.

Strategy Tip: Automate QA testing to reduce squad dependencies.

Work debt management

It's natural that you would strive to create the best possible products and solutions. Unfortunately, it is often the case that limited resources ranging from budget restrictions to lack of time, team dependencies, or lack of specialists will make implementing the best possible solution impossible.

This is especially applicable when your organization is in the startup phase, where you will frequently run into limitations.

Most of your experts and specialists want to do their best possible work, so they will work towards creating the perfect solutions. That's why you hire them. You need their expertise in certain areas to improve your products, services, technology, etc.

As a result, you often get technical experts who want to develop the perfect technical architecture, capable of scaling to thousands of users or handling any kind of usage peaks. Or financial experts who want to automate the full billing cycle from end-to-end, so you can process thousands of transactions with checks and controls in place, so you don't need manual intervention. "Perfect" solutions like these are very good and may seem like a must-have, but they're often not. As a small organization, you often only have a few customers and your focus should be on building the basics and foundations of your company and products. Doing so to any level of quality within your limitations may require letting go of perfect solutions to simply create solutions. If you don't, you may end up with half a solution, rather than anything near a perfect solution.

It's difficult to choose to focus on a less-than-perfect solution or to skip using an implementation your team would like so that you can focus on more pressing needs and problems. It's a complex trade-off between short-term and long-term. Too much focus on short-term and you will compromise your future. Too much focus on long-term and you won't have the resources to deliver value now to make your business run.

Each time you make trade-off decisions compromising long-term solutions for short-term feasibility, you're creating a half solution that will leave work for later. This is known as work debt. Or, in technical environments, technical debt.

Work debt is not necessarily bad. Take the parallel of financial debt. For many people, buying a home is only possible with a bank loan, or assuming debt to a financial institution. If the mortgage is chosen correctly and monthly payments are not too high, taking on debt allows individuals to buy something much bigger than they could afford when they took out the mortgage.

Work debt, as well as financial debt, is just part of the game. Just keep in mind you will have to pay it back. You need to have work debt so that you can create something bigger for the organization rather than using all your resources on one solution. It's not reasonable or feasible to develop your product by creating the final solution for each component or process in one go. Instead, you have to start small, make compromises, and choose solutions that best benefit short-term and long-term goals. However, it's crucial to create a work-debt backlog, so you know where you made those compromises and what you will need to improve later.

Taking on work debt is a normal Agile mindset approach. Why? If you aim to create the final solution all at once, it will take more time before you can deliver value to your users.

Fig.20. Delivering value earlier

Instead, you can use something of an evolutionary approach. This will allow you to create basic solutions which you can develop and deliver well in the now without wasting resources on massive implementations you don't currently have the resources to finish or the customers to properly utilize. Here, you can work on developing solutions that deliver value now, which you can improve later as more resources become available. You will have a large backlog to work on, but you will have more solid components and will be more able to deliver a solution to the customer.

Unlike financial debt, work debt is owed to no one but yourself. You will have to go back and finish improvements to create perfect solutions later. Achieving this means creating a backlog so you can track what needs improvement, how you intend to get there, and why. However, you always the possibility that your strategy, ambitions, or long-term goals will change and your debt backlog will become irrelevant.

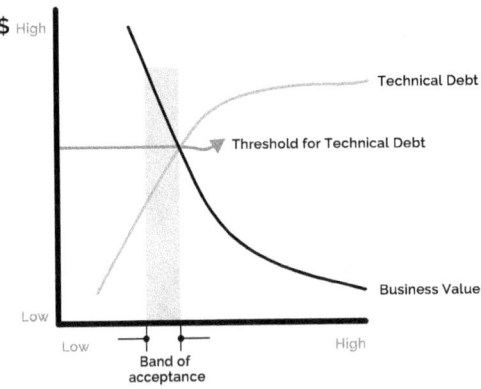

Fig.21. Work debt management

Strategy Tip: Keep track of work debt in a backlog so you can easily go back and make improvements.

My advice is that you keep track of work and technical debt you create, but don't let it get out of hand. When work debt is too big, you will compromise quality. You then have to slow down to reduce work debt and improve existing features before going back to developing new features.

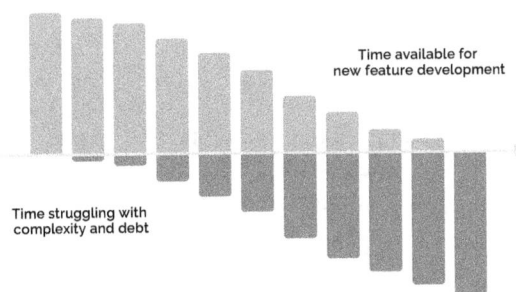

Fig.22. Measuring work debt

Most importantly, work debt can become one of your biggest bottlenecks. If your company and customer base are growing, you will have to grow processes with it. If you imagine that your debt backlog includes automating manual processes for onboarding new customers, it will eventually catch up with you. Assuming your team capacity is constant, you will eventually reach a point where you spend so much manual time implementing customers that your team no longer has time to develop an automation solution. You would need an intervention to reduce work debt and get things back on track.

Track how much time and resources you spend on not having final solutions. When it takes more resources to work without the final solution, your work debt is getting out of hand.

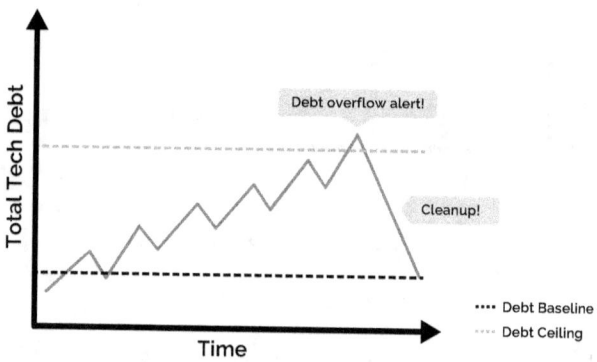

Fig.23. Measuring technical debt

Strategy Tip: Don't let work debt grow out of hand.

Conclusion

People drive your organization. Their work delivers value to the customer, whether with new features, documentation, or any other type of asset. Defining that work and how it should be completed is often crucial to achieving value, especially as companies scale and begin to take on more and more members.

The challenge is that in a scale-up company, teams grow quickly and are often still in consolidation stages, so work is not always defined. At the same time, teams need structure connected to vision in order to effectively deliver value.

Defining and standardizing work and processes across your organization will help teams drive value. Here, you can identify bottlenecks teams are facing, build priority lists, and assign specific roles and processes across the organization. These solutions should be long-term to ensure teams can move forward with solid structure and processes.

Process Management

"Many of our procedures were not designed at all, they just happened"

– Michael Hammer

Process management is the scope in operational management where you define how - how you will deliver work and how you will translate your organization's vision into value.

Designing internal processes is a key role of operations management. However, it is a process that will change dramatically depending on your organization and even individual teams. Numerous frameworks and methodologies exist to help plan and structure processes, how work is approached, and how it is completed, but you must understand what you need and why before making a choice.

Many frameworks are not a complete solution. They're designed to create a structure to improve the efficiency and efficacy of processes. They most likely will not perfectly fit your organization. Some teams inside your organization may require different frameworks or solutions. In this chapter, I'll discuss process management options and solutions as well as the factors playing into when and how each solution could fit into your organization and operational model.

Process design and management typically begins with high-level definitions, moves into roles and responsibilities, and then tackles actual tasks, task order, and process iteration.

Creating a process in this way allows you to set and define goals, what needs to be done to reach those goals, who should be doing what, and then create a logical order to complete those tasks, so that your process directly contributes to value. Good processes are designed around the high-level goal of value, with tasks and people organized in layers under that.

Processes drive your organization. You won't need documentation for every task, but well-designed processes will bring complex operations together to ensure efficiency and continued contribution to the value stream. At the same time, processes play an intrinsic role in your operations. Understanding high level frameworks as well as the in-depth steps of creating a structure and framework for those processes and team management will help you to create solutions that benefit your organization.

This chapter is structured to move through a high-level view of operations frameworks such as Agile and Lean, into roles and responsibilities, and then into detailed and practical process design.

Agile vs waterfall

"Waterfall" is a common form of work and project management, consisting of several distinct phases. In waterfall, each phase moves sequentially to the next as the previous is complete. You cannot return to a previous phase, and everything is conducted linearly and hierarchically. While previously the most common form of project management, waterfall is quickly being replaced by Agile.

In the waterfall method, development has a rigid sequence where you create a plan, implement, test, and deploy software. This model works well when the plan is perfect and there are no issues during that sequence.

However, the reality is that software development is often not perfect. It's sometimes impossible to create a perfect plan, simply because you are forced to make assumptions at the start. Many issues that occur during development are only discovered during development.

Agile is designed to overcome this precise problem by breaking a large plan into several smaller iterations. This enables teams to discover problems early on, deliver value more quickly, and adapt or change the end-goal as assumptions are proved wrong or new goals are set.

WATERFALL

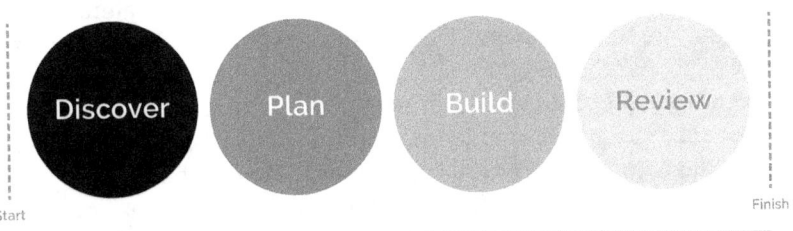

AGILE

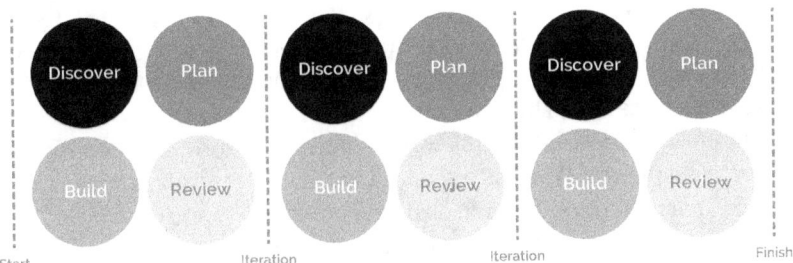

Fig.24. Agile vs Waterfall

This helps your organization to deliver value sooner because it can deliver smaller iterations of the software more quickly and with less risk.

Moving from Waterfall to Agile requires a huge shift in mindset because organizations must change not only how they work but also how they think – shifting focus from performing a process well to delivering software. No organization exists to perform a process well, which is why I firmly believe Agile offers numerous advantages. Waterfall organizations also typically work in silos, separating core processes and teams. Therefore, moving from waterfall to agile will require significant changes across the organization.

The easiest way to visualize the differences between Waterfall and Agile is with an analogy. I like to use travel. Here, I liken waterfall to a space expedition while Agile is more like a trip to the grocery store. How does that work?

In Waterfall, any process is thoroughly planned, scheduled, and worked out to the tiniest detail. The teams involved will likely spend months considering details. They will focus all resources on diminishing and predicting obstacles and possible scenarios to create mitigating measures. This high level of preparation means that there is no Plan B. All available funds must be spent on one route. Minimizing obstacles and problems is the highest priority. Once the journey to "Space" begins, there is no turning back. Teams must stay on track and use thorough documentation to avoid unexpected scenarios and to protect the outcome with clear negotiation instead of collaboration.

Staying on track ensures that development teams do not allow the product to shift course. This approach is designed to mitigate risks, but, naturally, creates many. Software companies that choose a single course with no opportunity for change might risk their product failing due to unexpected obstacles. If the journey becomes too expensive or the end goal is too far away, your team might miss the ability to launch a product onto the market, in the same way a journey to space might result in a missed opportunity to land on the moon.

Agile is much more like a trip to the supermarket. Every week or month you plan to go to the supermarket with the intent of quickly and efficiently completing a chore. It's not something you put a lot of time into, you just do it. On your way there, you might encounter obstacles such as a detour, traffic jams, or even a roadblock, but you'll quickly work your way through them and get to the store anyway. If you'd planned for these obstacles in advance, you would have spent a great deal more time planning for those and other obstacles, distracting yourself and your resources from the real goal.

Agile frameworks work in this way because they don't focus on planning, they focus on the goal. Work is organized into short intervals, which allow you to quickly change course to avoid obstacles as you go, so you can take a flexible approach to how you are arriving at your destination.

Making the shift between agile and waterfall requires a deep change in mindset. When you want to run Agile operations, you have to either hire people with that mindset or train people to work in an Agile way. Asking waterfall-minded people to work in an Agile way is a difficult process, and one where every small step is a huge barrier.

For example, waterfall-minded people prefer working in large batches because it's more efficient, like factory work. In theory, it's more effective to make more changes to a software module because you avoid the overhead of testing, deployment, etc. While this is more effective in theory, users do not receive new value during the process. In addition, you increase risks by skipping small steps with validation, which means that you may spend hours on a solution that doesn't work for the user. Switching to an Agile mindset and taking small steps with frequent validation will allow you to reduce risk, deliver value more quickly, and ensure that the end-product is what the customer really wants.

Waterfall-minded people also break epics down in different ways. The waterfall method would naturally result in breaking an epic into processes, "create functional design", "develop", "test", "release". An Agile-minded person would break the epic into smaller releasable features that has the end-user in mind and directly contributes value; "Signup form", "send email confirmation", "Connection with back-end CRM", etc.

Making this shift requires an immense amount of effort and coaching to change how people think. In my experience, it is significantly more cost-effective to align your recruitment with Agile and hire Agile-minded persons from the start to prevent this issue. However, you may still have to coach waterfall-minded people in your organization. If you are running Agile or switching from waterfall to Agile, it is important to be aware of these people, to offer development opportunities, and to ensure they can work in an agile way.

At Nmbrs, Agile is a significant part of our daily development. Agile's core manifesto of embracing change, delivering frequently, maintaining a close connection between teams and stakeholders, generating close feedback loops, generating continuous conversations and constructive feedback, and promoting sustainable development are key to what we do and how we are able to do it. With no specific need to plan for every obstacle, teams are able to self-organize, fine-run, and adjust in short intervals to complete work and reach goals as efficiently as possible. When obstacles occur, teams simply have to redesign the short iteration, be agile, and take small steps to move in the right direction.

Strategy Tip: Ensure that everyone adopts Agile and offer coaching where necessary.

Agile and Lean methodologies

It's important to define your approach to software development before going into the details of process design. The way you approach software development greatly impacts the way you design and implement processes, as well as the value stream as a whole. Here, Agile and Lean are two of the most common approaches for software companies.

Any organization's end-goal is likely delivering the best quality or the most value, and as quickly as possible. There are many tried and tested methodologies that will help you get there. Most of these methodologies trace their origins back to the 1950s, when Japan began its economic recovery from the second world war. Japanese businesses looked to improve processes and eliminate waste to reduce costs, so they could more easily compete in international markets.

The most notable of these businesses was Toyota, which developed the Toyota Way and the Toyota Production System (TPS), integrating a socio-technical system with well-defined principles to improve the efficiency and quality of their manufacturing and logistics systems. Their methodology developed around just-in-time production and waste-reduction, with principles and philosophies based on Japanese principles of Kaizen (continuous improvement), Genchi Genbutsu (go and see), Jidoka (fix problems on root causes), Heijunka (even out workload), and respect and teamwork. These principles were implemented from top-level management to the work floor and shared with partners and suppliers.

During the 1990s, this methodology gave birth to Lean manufacturing or Lean production. Lean is based on waste reduction, with the principle that anything not adding value is waste and should be reduced. Lean works to enhance productivity by simplifying operational structure to make performance and management easier.

Lean is primarily used in manufacturing and production industries, but its principles and philosophies apply very well to software development, which, at a certain level, is also a production system.

At the same time, many practices and methodologies have been developed and validated around the concept of improving software quality and the success rate of software projects. Iterative and incremental software dates to the 1960s. However, it wasn't until the 1990s that many of the lightweight methodologies used today began to emerge. Rapid Application Development, Scrum, Extreme Programming, and many others emerged as the software boom forced companies to economize to remain competitive.

In 2001, a group of 17 software developers met to discuss various methodologies and, bringing their ideas together, wrote an outline for a new methodology. The Agile Manifesto brought together ideas from Scrum, Kanban, and others – defining a software development methodology focused on creating end-value instead of rigid processes.

Agile's basic principles include:
- Individuals and Interactions over processes and tools
- Working Software over comprehensive documentation
- Customer Collaboration over contract negotiation
- Responding to Change over following a plan

The Agile methodology was revolutionary when compared to the waterfall model which was then the standard.

It's important to note that there are other approaches which may be more suitable for your organization depending on your environment. However, no methodology or choice will perfectly fit every aspect of your organization. For example, teams working with Agile frameworks experience Agile's many shortcomings including reduced predictability, greater demand on developers and clients, and a reduced focus on documentation. Depending on your organization and its focus, this approach may not work for you.

Agile and Lean are strongly connected in that Lean is an Agile technique. While many large corporations run Lean through Six Sigma, fewer and fewer software companies are doing the same. There is a reason for this, and it ties into how each affect organizations at an operational level.

At the same time, many COO's are unsure which to bring into their companies. With large-scale companies often running Six Sigma, it can be tempting to incorporate Six Sigma and hire a Black Belt instead of an Agile Coach.

There are pros and cons of each, and a natural overlap in some aspects of each, but there are specific considerations for most software companies.

Both frameworks allow you to organize systems and processes to more quickly create value, but with a few key differences.

Lean's focus is on reducing waste and eliminating unnecessary complication. This means simplifying documentation, reducing time spent building structure and software that you don't need in the now, and reducing anything that adds more work now without a direct and guaranteed contribution to value. Six Sigma was developed to reduce defects in the production process, based on the concept of defining, measuring, analyzing, designing, and verifying any new process to create incremental improvement over time. It's specifically designed around production and was developed from the Toyota production system and implemented by Motorola, General Electric, and others.

Agile is much more about empowering individuals and teams to take the full production process into their own hands. The idea is that an individual can then perform tasks in a way that makes sense based on the situation. Rather than strict control of process management to reduce waste, teams are given more control over every step of the process so they can innovate, respond to situational differences, and make changes. Rather than a single Black Belt to oversee changes, each team is responsible for their own quality and output – empowering them to succeed or fail on their own terms.

While both work to improve quality and output, Six Sigma does so through rigid processes and reducing risk while Agile does so by innovating and moving quickly in small steps to create large improvements. As a software company, I believe you need Agile rather than Six Sigma.

Today, many software companies use and implement Agile for a variety of reasons. While it is not perfect, I believe the Agile principle promotes an organizational mindset that enables companies to successfully deliver software applications. Agile is especially relevant where today's ever-changing environment requires a mindset of continuous learning, adaptation, and discovery, such as for product requirements, technologies, business model pivots, etc. This matches the concept of the infinite game, where delivering value is a journey and not a destination. For this reason, the processes in my Vision to Value Framework are based on Agile principles.

> **Strategy Tip:** I recommend using Agile, because it empowers teams and puts quality control in the hands of every member of your teams, while Six Sigma puts it in the hands of one person per team.

Agile frameworks

Building the structure and processes for a team is a long process which requires trial and error and a deep understanding of how and why teams work. However, in most cases, you don't want to start from scratch. Building team processes from the ground up is extremely time consuming and, in most cases, not actually necessary. Dozens of agile frameworks exist on the Internet, and in books – compiling best practices and tested practices from other teams. You can adopt and implement them for your own organization with comparatively low effort and a relatively high payoff.

These frameworks offer numerous advantages in that you don't have to start from the ground up and you can leverage someone else's research and knowledge. You won't have to research or learn by trial and error. However, you will have to understand the framework and you will have to understand why you're implementing it. Agile frameworks come in many variations. Understanding the mindset and reasons behind those variations before implementing one will help you make a better choice.

Agile frameworks can have disadvantages as well. For example, if something isn't part of the framework you chose, you likely won't follow or implement it. This can be limiting because you might not try something that would be truly beneficial to even one team in your organization.

My advice here is that you spend a great deal of time on research, don't be afraid to look at numerous examples, and try to choose something that fits your teams. If you truly understand your framework and how it fits into your environment rather than implementing it because it's popular, you'll get a lot more from it. There are many Agile frameworks designed exclusively for small organizations, so you don't have to choose something geared towards enterprise.

When I started, there were no real adoptable frameworks. Today, that has changed and making use of available Agile frameworks is the de facto standard. As I said before, it's crucial that you take the time to research and choose a solution that works for your needs. Here, I have highlighted several Agile frameworks which you can use as examples. I would also like to recommend doing your own research to determine if there are better solutions for your needs.

LeSS[7] - Large-Scale Scrum or LeSS is an Agile framework incorporating the principles of Scrum into a scaled framework designed for large groups and scaling. With the core principles of Scrum (Balancing abstract principles and concrete practices), and a focus on creating solutions for implementation, LeSS provides a set of principles, experiments, and a concrete framework with which to structure teams and scale.

SAFE[8] - Scaled Agile Frameworks or SAFE is designed to address the problems of scaling beyond a single team while retaining Agile methodology. SAFE is one of the most popular Agile methodologies, but it does have limitations and you should review its specific applications. SAFE was first designed in 2007 and works to increase the scope and planning horizon of agile while introducing a top-down hierarchical environment. SAFE also offers multiple configurations, making it suitable for a range of businesses.

DAD[9] - Disciplined Agile Delivery or DAD builds on Scrum, Agile, and Lean to enable simplified process decisions. With a focus on people and learning, as well as an end-to-end approach, DAD moves beyond Scrum to support long-horizon goals and implementation as well as incremental and iterative solution development and delivery.

---

7   https://less.works/
8   https://www.scaledagileframework.com/
9   http://www.disciplinedagiledelivery.com/

There are numerous Agile frameworks available and most have advantages. Spend some time researching your options and choose something that best fits the culture, goals, and organization of the company.

Strategy Tip: Based your operational model on an existing framework.

Planned vs. Just-in-Time

Scrum and Kanban are two of the most popular Agile methodologies on the market. Many organizations want to know if their teams should run on Scrum or Kanban, but it's important to understand that each methodology has its own goals and purpose. If you know what each is best suited for, you can likely easily match either methodology to your teams or team. The results will not be the same across your entire organization.

Kanban is essentially a framework built around continuous development and delivery. Here, teams consistently tackle small tasks using visual planning to track projects from start to completion.

Scrum is more about planning, where complex tasks are split into stories and visualized into a workflow. Sprints are planned periods where teams commit to a certain amount of work in a specific period, enabling consistent and long-term work delivery.

Scrum and Kanban each suit very different types of teams. If work is predictable and comes in cycles, it works very well with a scrum sprint. If it is more ad-hoc, a scrum sprint may not work at all and Kanban would be a much better solution.

I have often found that one methodology won't work across your entire organization. Instead, the methodology should be matched to your team's work requirements.

For example, at Nmbrs, our operational team Squad Infra receive a lot of internal requests regarding infrastructure. Using Scrum, they frequently felt as though they were failing because ad-hoc requests meant they were unable to plan work. I quickly realized their problem wasn't about planning, it was about methodology. Switching to Kanban gave the team much more freedom to simply visualize work and start without having to plan a sprint. At the same time, our teams working on features development benefit from Scrum because they can define clear sprint objectives, do proper estimations, and deliver with predictability.

Both Scrum and Kanban have pros and cons, but in most cases, it's not about which is better, it's about which is better for your teams.

Strategy Tip: Match methodology implementation to teamwork requirements.

From responsibilities to processes

Defining a process means organizing how work should be completed and why. However, once you define process documentation, you still have to outline responsibilities, which will help you to distribute work across teams. Processes and responsibilities are intertwined because processes must be defined within the framework of roles and responsibilities.

The question for many organizations is, which should be defined first? Processes or procedures and responsibilities?

It's better to begin with responsibilities and then map them into procedures across your organization. These responsibilities can then be grouped into a role and roles can be grouped into a function, which you can use to define processes. Responsibilities detail what and why something is being done, and processes detail how actions are being performed, so it makes sense to define the "What" before going into "how".

It's incredibly important to define responsibilities as part of business processes. If you assign processes without responsibilities, people will simply perform the process without an end-goal or understanding why, which will not contribute to your value stream. Assigning responsibilities avoids the issue where people perform processes by route, just going through the steps to get to the end, without paying attention to who is performing responsibilities or why.

Here, you can take one or several roles, define responsibilities for them, and organize those responsibilities with processes. This works in layers, which you can define by stacking responsibilities inside of processes inside of roles.

- Responsibility (code quality)
- Process (Perform a code review - someone has to open code and review quality against guidelines)
- Role (Who is performing this role)

If your responsibilities are properly defined and distributed across teams inside of functions, you won't actually need as many processes. When people know what has to be done and who has to do it, they can perform those tasks in whichever way is most applicable and logical for the environment. It doesn't often matter how responsibilities are completed so long as they are completed. The experts you hired to work know how to do so better than you (which is why you hire them).

Processes should normally be an extension of responsibilities. I normally find that it only becomes truly relevant to design a process when there are external stakeholders involved or the work is critical and complex. Process design becomes more valuable in situations such as when an external stakeholder needs to know how a certain activity is handled or you find a certain activity requires so much coordination you want to clarify who is doing what. Similarly, process design is invaluable when complex sequences such as product updates must be consistently recreated in the same way and to the same standards.

Process design is only relevant when it directly adds value to the process, rather than simply highlighting logical steps or actions inside of responsibilities. If people know their responsibilities, they shouldn't need processes for routine, simple, or very linear tasks that do not require a great deal in terms of compliance, auditing, or quality assurance.

At Nmbrs, we implemented process design for critical activities requiring a great deal of coordination, which benefited from optimization. For example, product updates and customer support. Everyone was already aware of their responsibilities, but by designing processes in collaboration with experts, we were able to determine what to do in edge cases and which role(s) would be taking on certain decisions. Once the processes were described, we were able to improve our development and support services, as it was clearer what had to be done and by whom. In addition, it improved the onboarding process for new colleagues because they could simply read diagrams and easily view their tasks and direct colleagues.

Processes are intertwined with responsibilities, so any activities in a process must be linked with responsibilities. At the same time, responsibilities must be broader than the processes the role is linked to.

Strategy Tip: Establish clear responsibilities before designing processes.

Processes design and implementation

Process design is a crucial element of operations. Defined processes allow you to more easily share how work is completed. This, in turn, benefits onboarding, quality assurance, performance monitoring, and external process management, or performance review such as auditing and compliance.

There are many formal languages to help you describe processes including BPML, BPEL, XPDL, and many others. These languages work to create structure for processes with designed workflows, process modeling, and workflow implementation.

While formal languages can be useful in describing processes, I personally find that they are too much for most tech companies. On the one hand, structure is a good thing. On the other hand, rigidly formatted work processes can disrupt an agile environment. Most business process languages add value to very large organizations requiring externally formatted processes for the purpose of external auditing. I believe process languages are overkill for smaller scale-up organizations. They extend well-beyond the needs of small tech companies. A lighter approach is much better suited and will offer more value.

Agile teams are often empowered to design their own processes, which creates a deeper connection between the team (the people who run the process) and the work being completed. However, teams often have very little understanding of BPM or the deeper formal requirements of processes such as auditing, segregation of duties, compliance, etc. I personally believe you need BPM as well as Agile.

At first glance, it may seem that BPM is not at all related to Agile methodology. After all, BPM is about defining structure through formally describing business processes, giving teams a well-established workflow and process to follow. It's often applied in environments with little discovery or exploration, where the focus is to produce more and improve output by maintaining high quality standards. This type of documentation is handled by a BPM consultant and rarely used in day-to-day work. With BPM, processes are often very well outlined and written down, but rarely actually used or referred to by the teams who should be using them.

Agile contrasts this with iterative work methodology, where retrospectives are used after work sprints to keep teams learning. Every workflow might be different as improvements, collaborations, and other aspects are introduced to improve productivity. Team collaboration, engagement, and constant adaptation are key to Agile processes.

How can these two vastly different methodologies work together? In my experience, the risk for process design is that if it is done in an Agile way, with no company-wide standards, everyone will build processes in a different way and there will be no set standards teams can use to share work using the same processes. It's important to have a common tool or standard everyone can use to design a process.

Standardizing tools across your organization will help you tackle problems including the creation of valuable processes and giving teams individual process ownership. This will allow teams to create their own processes, change them when needed, and stay in control of their own work environment. It is highly advantageous over having an external designer or business analyst own the process because it will help you tackle two of the largest issues relating to process design and implementation; ensuring individuals doing the work understand the processes and ensuring they follow processes.

Asking teams to aid in the process design with the support of a BPM expert will help you to bridge the gap, creating formally valuable processes which also works for the team implementing them. It's crucial to avoid the old-school design process where you have a business analyst pushing a process down because teams won't truly accept and use a process unless they are involved in its creation. The designer should motivate the teams and people involved to see the value of the process, which ties into another important step – why does the process exist?

Not every team or task needs a process. If you overdesign and have too many processes, they will get in the way of productivity because people spend too much time checking and following process. Any process you have should add value, it shouldn't be a simple common-sense task list for a series of actions. You can balance this with BPM by giving teams the opportunity to design and review their own processes to validate and ensure value.

Many organizations will implement processes for compliance. You need processes to share work methodology, activity, and how you implement change-management internally. You also need processes for quality management, so that you know how work items are handled and the steps you take to achieve them. Here, processes also help with onboarding, because you can easily share how work is accomplished in the company or the role because it is a documented process.

However, my experience is that if you create valuable processes which can be used by your teams to improve work and efficiency, they can later serve a secondary and valuable purpose for compliance. Designing processes solely for the purpose of compliance will result in process documentation that adds little value and that may actually get in the way of your value stream. Therefore, my recommendation is to design processes for teams to improve work and to use formal functions as a secondary element of process design.

You also have to connect functions/roles to processes, tying what must be done to who has to do it. So, if someone is in a certain role, they can see which process they are part of. This is relevant for individuals understanding their own roles and contributions as well as onboarding and understanding the scope and responsibility inherent in a role.

Good process design means outlining processes and functions, so that you can define high-level activities and then zoom in to more detail. This will help you define and understand what you want to map as a process or procedure. Here, a process is a series of related tasks or methods that generate output. It consists of an input such as materials, code, information, etc., a process or work method and sub-processes, and an output, such as a new application feature.

A procedure goes into more detail, describing who in the team is responsible for each part of the process, when each part of the process occurs, and the specifications applicable to each part of the process.

It's also important to align any process with long-term value and sustainable flows. Once a process is described, it should align with a long-term strategy. You shouldn't define a process to achieve something more quickly if it's not directly contributing to long-term value. A bad process can backfire because it can help you do unnecessary work more efficiently and can take up resources which would be better utilized elsewhere.

At Nmbrs, we created a simple solution in the form of the Nmbrs Book, an internal platform where teams can access information related to processes. Here, any individual can review business processes related to their role as well as to specific value streams and activities. Each process is designed to allow anyone to view their responsibilities based on role, to expand each process to view more information on activities and tasks inside the process, and to create a complete and logical structure for taking the process from start to finish, including handing it off between teams. Making work available in this very transparent way ensures that anyone and everyone knows what they are doing, when and what tasks become their responsibility, and what should happen at any stage of a project to move it forward.

This content is not static because it reflects the way the organization works.

It implements iterative improvement from Agile, which makes our organization very adaptable to change. Every squad member is a contributor to the Nmbrs Book, everyone is responsible for knowing and understanding the content and processes, and for checking for consistency. In this way, we make processes part of work, not an accessory to them. Processes are part of everyone's daily work and are therefore managed under the same Agile laws of and improvement cycles, allowing us to integrate the valuable formal definitions of BPM with the flexibility and continued improvement of Agile.

Strategy Tip: Involve your teams in process design.

The problem of processes documentation

Processes are typically defined in the form of a diagram or document tooling. While documentation is valuable for use in formal processes such as auditing and compliance, it's less valuable for work, because most people won't ever read or use it. Diagrams and documents are often difficult and outdated, offering little value to teams.

If you want teams to use process documents, you have to push and force engagement and continued use.

Teams should be able to follow the process without accessing documentation because the process should be a natural part of work. If it isn't, the process isn't working properly and it isn't useful.

Why? If something is useful, there should be triggers inside the organization, team, or unit to keep it working because it adds value. If something adds value, teams will use it naturally as part of their workflow because it makes work easier, more productive, or otherwise better in some way.

You can encourage adoption of processes by keeping documentation light and high-level. Process documentation should incorporate into tools and workflows, not necessarily on paper. You will need paper or written process documents for compliance purposes, but you don't need a great deal of detail for this reason. The lighter you keep processes, the better and more useful they will be, and the less you will have to update them.

When you implement, you can do so in the form of continuous integration and tooling, which often allow you to directly integrate processes. For example, if you can force certain workflows in a tool, people don't have to understand the process, they just have to work with the tool to automatically follow the process. Of course, some tools don't support workflows. In this case, you will have to spend more time agreeing on a process with individuals and teams.

Documentation works best as a sort of blueprint, so that you know what you want to have and why. However, you don't have to write long or complicated process documentation, because, chances are, no one will read it.

One example of this approach is to model processes inside your project management system. Here, you create workflows mapped to the steps in your process, so they are automatically part of any project. At Nmbrs, when teams work with "product councils", steps inside the project management system include "preparing for council", "ready for council", "approved", "In progress", "on hold", and "shipped". This helps us determine whether a ticket was tested by the right group, who selected work items, and who completed work items. It's very useful for auditing, because the tool tracks and registers the steps of your process, so you can always see when work does or does not comply.

Strategy Tip: Implement processes into tooling in addition to documentation.

Measuring Process performance

In any instance where you design and implement a process, the end-goal is to add value to the organization. Whether through improved efficiency, better communication, or by aligning teams with the same work-methods, processes must contribute value. Process monitoring is an important part of implementation because it allows you to define and track what contributing value looks like in your organization.

If you design a process, you have to be able to measure its efficiency and efficacy. Here, you have to explicate relevant metrics per process, including quality, what measuring value looks like, and key performance indicators. This also means understanding which KPIs are relevant for each process so that you can define what a good process looks like in each environment.

For example, if you take a process outlining how you handle customer support incidents, some relevant KPIs would include the volume of incoming tickets per channel (email/phone), the number of people or FTEs, ratio of tickets per customer, and so on. This data would enable you to monitor the performance of the process so that you can measure improvements as you tweak how activities are performed.

It's also very useful to link a view to see logs, work items, or anything related to the process next to the KPIs. For example, if you have a security incident process, it's useful to have a link to security incident logs, so that you can see whether they were handled properly.

Strategy Tip: Define relevant KPIs and link views on your processes.

Process automation

Once you've created processes, automating those processes may seem like a logical next step. However, while robotic process automation is often easy and profitable, it's equally as important to define a process for automation. This process should include manually validating and ensuring the quality and efficiency of a process before automation.

Why? Process automation can help you to perform a process more quickly, more often, and more repetitively. But it can't help you improve the process, it won't help you determine if you need the process at all, and it won't review for mistakes. If you automate a faulty or unnecessary process, you're performing something that is necessary or wrong over and over again, without the benefit of a manual quality check or review. This can backfire in a big way, because small mistakes can become significantly larger when automated due to the simple process of repetition.

Automation can serve a valuable purpose in that it will save labor and improve processes by increasing efficiency and speed. However, implementing automation in a way that directly benefits your business necessitates having a good understanding of your processes, the value they contribute, and what automation will enable.

Even long-established businesses sometimes automate every process they own. They see a return on value and assume success, but without an understanding of their processes, miss out on large opportunities to optimize in other ways.

Taking the time to map, analyze, and validate your processes before automation will help you to review your processes for quality, efficiency, and efficacy. For example, many existing business processes are too complex for the purpose of automation, which can be helpful but is often unnecessary for Agile teams and unnecessary for quality assurance or compliance. Simplifying processes before automation improves your ability to automate them, while improving their implementation inside your organization. Automation may also make several parts or steps of your processes redundant. For example, if you have checks in place to check for human error, automation will make those steps obsolete because machines don't make those mistakes.

You should know your processes inside out before automation. The people using them should understand how to use those processes, what they are for, and should follow them. In addition, you should validate those processes, challenge them and improve them before automation, because once you automate, chances are, the lack of manual intervention will mean that improvements and iteration will stop or slow significantly.

> **Strategy Tip:** Manually run your processes and validate them before automating.

Follow the rules to break the rules - Shuari

In any work environment, there are instances when individuals and teams can bring change to an organization, can choose to follow the rules, or choose to improvise. The question is always, "when is each appropriate and how or why does it add value to the organization".

For example, when someone new starts in your organization. This person is highly motivated to show their value and to bring their skills and experience into play so they can begin contributing as quickly as possible. The environment is new and there are many unknowns, such as processes, culture, people, etc. As a result, the new person's instinct is often to overlook existing processes and work methods and begin changing things to what they are accustomed to, to make it more comfortable.

In an example; if you've hired a new senior developer and she starts working on one of your products but doesn't take the time or effort to fully understand what was developed before her. She begins introducing changes because she believes the current code base is difficult to understand and too messy. At what moment would it be correct to implement change instead of following existing processes?

It's often important for people to be critical of their environment so they can change and improve it. But it's also important to understand how existing processes work, why they are in place, and how they add value. Here, I believe the Shuhari principle defines a process of learning, mastering, and then improvising and improving which in making this distinction.

While common in martial arts, it is good advice for gaining mastery in the business world as well. The concept translates into "To keep, to fall, to break away."

The concept incorporates 3 stages, Shu, Ha, and Ri.
- Shu - "Protect" or "Obey - here, you learn the fundamentals, traditional wisdom, and techniques
- Ha - "detach" or "Digress" - here, you break with tradition
- Ri - "Leave" or "separate" - transcend the traditional

The idea of Shuhari is quite simply that you must learn the fundamentals and basics of what you are doing before you can experiment and grow with it. During the Shu phase, your goal should be to learn as much as you can, follow traditional wisdom, and build something that works while you learn. Once you move on, you can begin to break the rules because you have the basic knowledge and experience to understand what will work and succeed.

First, you learn, run it with excellence, and then improve.

Aikido master Endō Seishirō shihan says:

"It is known that, when we learn or train in something, we pass through the stages of shu, ha, and ri. These stages are explained as follows. In shu, we repeat the forms and discipline ourselves so that our bodies absorb the forms that our forebears created. We remain faithful to these forms with no deviation. Next, in the stage of ha, once we have disciplined ourselves to acquire the forms and movements, we make innovations. In this process the forms may be broken and discarded. Finally, in ri, we completely depart from the forms, open the door to creative technique, and arrive in a place where we act in accordance with what our heart/mind desires, unhindered while not overstepping laws."

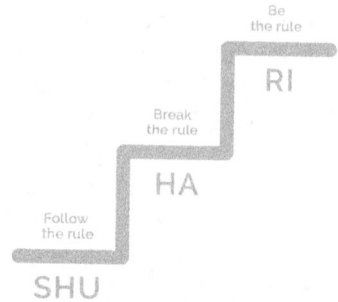

Fig.25. Shuari steps

When someone who knows your organization breaks the rules, it normally means something was impeding productivity. You should create a standard process during retrospectives to identify when and why rules are being broken. This information can be used to create better processes. When you do so, it's important to involve all relevant stakeholders rather than just technical teams. Why? A compliance officer might have very valid reasons for seemingly unnecessary processes or steps that might not be obvious to technical teams. Collaborating with all stakeholders on any process change will help you to improve how people work and increase productivity, without impeding other goals.

> **Strategy Tip:** Learn the fundamentals and create excellent products or processes before experimenting or changing them.

Introducing new methodologies

One of the most challenging aspects of implementing a new work process, whether Agile, Lean, etc., is not integrating it, it is ensuring that people understand not only the tools of the methodology but also its core principles. Fostering this understanding requires not only training and coaching on the job but also driving engagement and excitement for the methodology, so that people understand why they are using it and what they get out of it.

Sometimes, new work methodologies can be incomprehensible. Imagine a developer who is accustomed to working with a waterfall method. Her primary task is to technically implement a function design handed over by a functional analyst. Once completed, she would simply send it to a Quality Assurance Team. When you move this developer into an Agile work environment, she would have to participate in the functional design stage and QA, while retaining her technical role. The reaction is quite often, "I don't think I can contribute here, I am a coder"

It's important that people understand the purpose and responsibilities of their role, so that they can connect the scope beyond just coding, testing, or any other specialization. Connecting how they work across products will help you to achieve that.

It's important to remain watchful when integrating new methodologies to ensure they are adopted. Retain coaching and continue on-the-job training until work is flowing according to your operational model.

Practical workshops in real-life can be much more effective than theoretical lessons on the methodology. Most teams like to see how their work productivity can be improved with the new methodology.

The more accustomed your employees are to their old methodology, the more difficulty they will have in adapting to the new one. If your teams have used a single methodology for some time, you will have to invest a great deal in coaching and changing those habits. However, it will pay off as teams adapt to Agile or Lean and are able to work in more efficient and collaborative ways.

Strategy Tip: Use real-world coaching to drive engagement and adoption for new methodologies.

Creating a fail-safe environment

Most businesses don't want to fail. Instead, they create environments designed to prevent failure as much as possible. This is natural because "total" failure is an impossible-to-sustain business model. However, Agile environments enable you to take a different approach, "Fail Fast and Forward".

Here teams take multiple small steps forward in which failure is allowed, encouraged, and accounted for. This creates a failsafe environment where failure is inevitable and part of moving forward. New ideas, processes, and tools are tested incrementally to reduce risks, while allowing teams to grow and move forward. This process is difficult for individuals from strict Lean waste-management environments to adapt to, but can be extremely important for SaaS companies, where forward momentum is crucial to continued success.

Large-scale companies have traditionally attempted to mitigate risk as much as possible, but this inhibits growth. Some of the largest and fastest growing companies on the planet, such as Google, implement "Fail Fast and Forward" mentalities. Google's system encourages risk-mitigation strategies like A/B testing and backup plans but doesn't encourage absolute failure as an option. If something isn't working, you redesign, reimplement, and try again. These initiatives have resulted in numerous failures, but also in some of Google's largest products (Gmail, for example).[10] Google's 8 pillars of innovation provides structure for this innovation, creating risk mitigation, while leaving massive room for growth in any direction inspiration takes the company.

As COO, it's crucial that you create structure supporting failure, so that you don't only build processes with the mindset of moving forward, but also with room to fail, step back, and try again as part of a learning process. Failure is a

---

10 https://www.thinkwithgoogle.com/marketing-resources/8-pillars-of-innovation/

part of moving forward, and you need to build structure and support for it.

The easiest way to make room for failure is to make room within promises to customer and with deadlines, so that you have room to experiment and try again. Rather than rushing to meet output needs, you can experiment and create rather than simply executing a promised solution.

You can also give teams more room to be creative by communicating that failure is expected and okay and giving teams access to the problems they are solving. Handing creative and technical individuals problems rather than solutions to build will enable them to build better solutions. Your managers and leaders can contribute here by communicating and giving feedback in a constructive way without criticizing failure as a problem, only as an aspect of reaching a solution.

At the same time, this process will help you to build internal trust with teams, so that they feel safe trying new things.

Strategy Tip: Make room for small failures so that you can develop a culture of experimentation and forward momentum.

Processes and Compliance

Risk mitigation is one of the most important reasons to implement processes. Depending on your company's organization, this can be more or less relevant – e.g., if you're a startup, risk management is not immediately relevant – but it is something you should consider no matter what your growth stage.

Once you begin to scale, risk management is crucial to growth. For example, if you're operating a B2B company, your customers will look at how your company is running and operating. Other businesses need to be able to trust their partners and services, so they need to know you're operating in a healthy way. Having defined processes in place will help you to make those connections, because your business operates in a very transparent way.

Organizing and documenting your processes will also make it easier for you to apply for and maintain compliance and risk management standards. For example, should you need certification standards such as ISO, SOC, or ISAE3402, existing process documentation means you'll already have everything in place to make application as effortless as possible.

If processes are well-described, you facilitate the auditing process, which you will eventually need.

You can also use processes for compliance purposes. Many types of companies can greatly benefit from certifications and compliance because they add value through external trust from customers. While many people approach compliance and certifications as an intimidating process, it often becomes relatively easy once you have internal processes structured and documented. This will eventually greatly benefit the company if you can benefit from compliance and certifications.

At Nmbrs, we began using processes for internal improvement, after which we began to take external compliance and standards into account as well. Here, our existing processes became extremely valuable because it allowed us to very easily show compliance.

At the same time, I believe that compliance should not be your primary reason for creating business processes. They are much more valuable for internal improvement and optimization and compliance should be secondary. Formalities cannot be your main driver, or your processes will not reflect what is valuable inside your organization.

> **Strategy Tip:** Utilize existing processes for compliance but focus new processes on completing work.

Conclusion

Process management will give you the tools to effectively deliver work so that it contributes to value in a measurable and optimizable way. There are numerous methodologies and frameworks available to structure work, and many of them only require moderate adjustments to meet specific needs.

Choosing the right framework and process management for your organization will allow you to control growth in a way that works for your business ethos, your product, and your customer base.

It's important to create standardized processes, which will allow for automation, easier onboarding, and for measuring output and compliance. At the same time, it's important to continue to adjust your processes, introduce new ideas, and continue to improve your framework.

Processes should drive benefits and increase productivity. When that's not the case, remove them.

People and structure

"If you want to go fast, go alone. If you want to go far, go together."
– African Proverb.

Team performance is the scope in the operational model where you define who-who will deliver work and who will you turn your organization's vision into value.

While your structure and processes will help guide your organization on its way, it is the people and teams inside your organization that will truly make or break your ability to deliver value to customers. People are the greatest resource inside any organization. People do the work, create new things and ideas, and bring your company vision to life.

Good people and team management requires organizing teams in a way that aligns with business drivers. It means creating processes that encompass roles and responsibilities. It also means building team structure so you can allocate responsibilities, rely on consistent performance, and prevent issues such as bottlenecks. In this chapter, I will go over standard team organizational models, inspiration, and my own experience with designing and structuring teams.

Team structure

When your organization first launches, chances are that roles and teams are poorly defined, if at all. In most cases, founders will take on whatever roles they can, often filling multiple roles in a smaller capacity. As your organization grows, new employees are brought on to fill specific needs. However, because people still have to take on numerous tasks just to get things done, roles remain poorly defined. One person may fill multiple roles in several teams, and, as a result, even teams may be poorly defined.

As you grow, adding definition to roles and teams becomes essential to organization and management as well as for monitoring team performance, tracking completed work, and even delegating work. If teams are unsure of what they are supposed to be doing or how they're supposed to it, they will complete work that doesn't contribute to the value stream. This eventually becomes unsustainable as roles overlap, you lose work visibility, and task delegation is met with bottlenecks and confusion.

Creating fully structured teams that are aware of individual roles, responsibilities, and scopes, will help to resolve this issue, so you can easily manage and optimize work output.

Agile is one common method to structure teams. It allows you to organize flexible teams who take ownership of what they build. The goal is to have teams that can work in a quality and efficient way, without strict management and process guidelines.

It's difficult to define team structure from scratch. However, there are thousands of existing frameworks and ideas you can use to organize your teams.

The approach I've found to work well in my own and other organizations is based on the Spotify model. It is interesting because they began with an Agile framework and adapted it to something new. They incorporated a cross-functional structure in which teams exist within teams to bring experts together across the organization. This framework functions well for small organizations but is proven to scale, as Spotify operates it for 300+ engineers. Scalability is one of the biggest challenges of designing team structure, so Spotify's example of an extremely scalable and flexible framework is a good one.

Spotify's model also works to solve another crucial problem inside Agile teams – isolation. While Agile teams are beneficial in that they give each team the tools and talent to take complete projects from start to finish, they often isolate experts and individuals. When everyone is isolated, they may not be able to share valuable ideas, workarounds, and tools across the organization.

You may have people working in the same room who create duplicate solutions, simply because they haven't had the opportunity to share. If a developer in one squad solves a problem but doesn't talk to developers outside of that team, another team may have to build a new solution for the same problem. The Spotify model helps to bridge this gap by connecting experts across your organization based on their discipline or specialty.

Spotify Product Development/Spotify Engineering culture combines the tenets of Scrum with those of Agile. Here, Spotify shifted Scrum Master to Agile Coach, renamed Scrum teams to squads, and created micro-units inside the organization, each with their own autonomy and leaders.

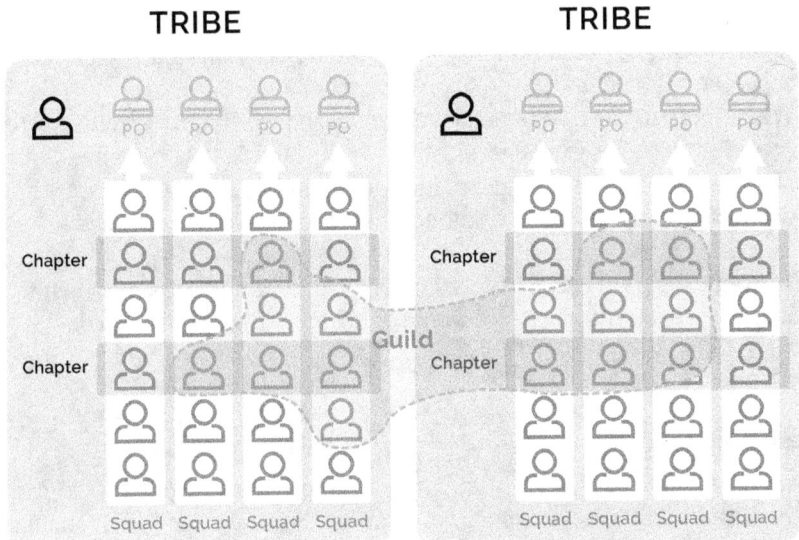

Fig.26. Team structure based on Spotify model

Squad - A small team owning part of a function, end-to-end, with a dedicated product owner who feeds the team products or stories to build. Squads have the resources to be autonomous, self-starting, and self-organizing. Renaming teams (to Squads), is one way to break up traditional ideas about what it means to be in a team. For example, traditional development teams only include developers, but a development squad might be much more relevant to the work being performed.

Tribe - A collection of squads inside the same business area. For example, a customer support Tribe might include developers, customer service represent-atives, and so on. Grouping individuals based on the business area they work in and linking them enables these groups of people to more effectively work together, even across teams.

Chapter - A chapter is another larger organization of individuals across teams, connecting people who work in specific areas. Your squad could be made up of people who work in front and backend applications. A Chapter might connect everyone who works in front-end applications, so that everyone can get together, discuss ideas, share insights, and work together to create a single product.

Guild - A guild is another large, cross-team structure, organizing individuals based on common interests. For example, you can create guilds around quality assurance, automation, design, or development. This allows individuals shar-ing a common work method or interest to share ideas, solve problems, and work together across the organization.

Using this sort of team structure with Agile allows you to benefit from Agile while reducing its drawbacks. By dividing people based on work rather than a core task like development, you create teams that own processes and take them from start to finish without bottlenecks. You also avoid isolating technical experts from others in their field. Creating structures to support cross-field interaction and collaboration also adds to efficiency and creativity because experts can share ideas and build on each other.

While this sort of Agile team structure doesn't work for everyone, it works very well in most software organizations, especially SaaS companies. Squads enable fast and lightweight product development for faster releases and iterations, because each squad can autonomously develop small releases.

> **Strategy Tip:** Don't start from scratch when designing a team structure, take inspiration from existing and working models and build or adapt from there.

Designing teams

In any case where you hire or bring someone into your organization, the primary goal is and should be achieving company goals and vision. Teams should be designed around the same purpose, because by designing teams and allocating resources in the right way, you ensure your organization has the resources (budget, manpower, software) to drive the organization in the desired direction.

Using organizational goals and objectives as a starting point for team design allows you to determine team size, which teams you need, and how to allocate tasks and responsibilities across teams. Considerations include designing teams around continuity, stability, and the value or importance of what the team is working on.

Your team structure should be based on how your organization prioritizes work and how value is defined in your organization. This ties into the provisional model and your market focus. For example, if your organization were to focus on Product Leadership, you would want to design development-oriented teams with a high focus on quality assurance, while leaving room for experimentation and innovation. If your organization were to focus on Operational excellence, you'd have to design teams with standardized processes for quality control, which will mean sacrificing freedom in terms of room for experimentation and flexibility.

At a certain point, my organization recognized that one of our organizational goals was to develop more HR features. Previously, HR feature development was spread across teams. We decided it should be more important and assigned 2 dedicated teams to the task. This helped reduce dependencies and overhead, because teams don't have to allocate work to each other. Instead, they manage communication together, sharing Agile ceremonies such as kickoffs, review sessions, and QAs, so they have less overhead.

This tactic can be valuable in numerous situations. Creating focused teams (instead of just development teams who work on anything) ensures everything is stable and important business goals are being attended to.

Focusing teams in this way also allows you to ensure that less important tasks aren't overdeveloped or overstaffed. For example, we have a mobile app that is important but not critical, so we have a very small team assigned to it. When we have a critical element, such as our payroll engine, which needs to be very stable, we ensure that those teams are bigger to ensure continuity. When people are sick or have to go on holiday, they can do so without affecting the team's ability to manage their routine tasks such as maintaining stability and updates.

The best way to prioritize team size, budget, and resources is to design teams around current and future development needs, work priority, and how the team delivers value.

Team design involves creating teams, assigning team purpose to teams, determining which teams you need where, how big they should be and why, how many chapters you have and so on.

Making these decisions regarding team scope and purpose can be difficult but it's important to remember that your organization will never be static. Teams always change their scope, size, and purpose to continue meeting the needs of the organization. A team may be designed to fit a specific purpose, such as working on specific modules/products, but that can change over time and they may eventually fill a completely different role inside the organization.

For this reason, I recommend designing teams around specific responsibilities rather than work items. Doing so means that when products and work items change or become obsolete, the team still has parameters and a scope in which to frame their work. When you use responsibilities to frame teams, you can also more easily change team purpose, without inherently changing what the team does.

In most cases, you want to develop autonomous teams that can develop solutions on their own. Designing good team structure will help you build focused teams that work towards achieving objectives rather than completing tasks, so they can proactively add value.

Finally, it's important to pay attention to how team members are managed and assigned. Good team design helps you set up teams that are the right size and scope for the business driver they are connected to, but organizations change over time. You might look up one day to realize that some teams are much too big for what they are trying to achieve while others are much too small. This happens naturally as business focus changes and a large team becomes obsolete. It can happen accidentally as you scale and HR begins to add new employees into teams without consulting team structure and design. In either case, it's important to create regular checkpoints to ensure teams still meet business goals and are still oriented and sized in a way that enables them to achieve their specific purpose and goals.

Launching a new team is an important moment and a key opportunity to align everyone with the purpose, scope, roles, and responsibilities of the team. My recommendation is that you always make time to create a kickoff for the team, where these factors are discussed by the team. I would also like to recommend creating new kickoffs when major changes happen, such as when several new team members are added or team purpose changes. It's always a good idea to create regular alignment meetings (say, every 6 months) to ensure everyone stays aligned. This will ensure that teams stay on track, remain aware of their purpose, and are better able to work towards their goals.

> **Strategy Tip:** Align team scope and capacity with relevant business drivers and objectives.

Empowering teams

Any organization designed around Agile will naturally be flat in comparison with one designed around waterfall or any other traditional framework. However, deliberately working to keep hierarchy as flat as possible will benefit your organization in many ways.

It's common for organizations to add on layers of management as they scale. Doing so creates layers for decision-making, communication, and hierarchy. This concentrates responsibilities such as hiring, firing, coaching, strategy, and planning into middle management. While historically very common, this sort of structure will impede team ownership and motivation. But, how do you empower teams in scaling companies while maintaining structure?

A flat hierarchy works to empower teams, giving them ownership of their own purpose and projects. Reducing the need for managerial approval removes bottlenecks, gives experts on teams the agency to make decisions quickly, and allows you to create a much more adaptive and flexible work and production environment.

Flat organizations are guided by their purpose and pushed by the empowerment of individuals. This means employees take ownership of their work and results in initiatives that will bring competitive advantages in dynamic markets.

Diluting managerial positions throughout roles and teams is one approach to achieve a flatter hierarchy. For example, the Product owner decides what to develop, the Scrum master is responsible for process, people coaches are responsible for personal development, and Chapters are responsible for guidelines and technical decisions. This approach allows you to allocate managerial responsibilities to experts in that domain.

So, in my opinion, the real key to flat hierarchy is creating a structure in which teams understand their purpose and are able to self-drive towards it, without needing layers to tell them what to do or how to get there.

Here, organizational paradigms are consistently moving towards flatter hierarchy. Today's hierarchies include everything from complete separation of thought and labor to completely flat organizations with Teal and Holacracy. In the latter two, organizations distribute autonomy across teams, operate without leaders, and distribute thinking based on skill, experience, demand, and willingness to contribute and take charge.

While flat hierarchies can seem extreme to those accustomed to management and leadership, they have been shown to be profitable. Organizations, especially startups, are increasingly going "flat" to improve agility, empower teams, and to put decision-making in the hands of experts, rather than managers.

Zappos first integrated Holacracy in 2014[11] to meet a need for greater individual autonomy and responsiveness inside customer service teams. The company integrated self-management to empower individuals and reduce the number of layers between employees and customers. Zappos still runs Holacracy, with flat hierarchy surviving an increase to over 1,500 employees, an acquisition by Amazon in 2009, and company growth from $184 million in gross sales in 2014 to $3 billion in 2017[12]. While it's unclear how much of this success is directly related to Holacracy, the company is just one of many which continue to implement and thrive with flat hierarchy.

The increasing trend towards flat hierarchy also reflects the world around us. Individuals are more often than not highly educated, highly motivated, and leaning towards self-management. Many work from home, in freelance roles, and are often highly self-motivated. With their own goals and ambitions, many people don't need managers to assign, distribute, and enforce work. Flat hierarchies enable you to empower people to work in a certain direction in whatever way is most possible and applicable. Many people will continue to rely on management to assign and distribute work, but culture and work are shifting and flat is becoming more and more common.

Developing a scalable chain of command with a flat hierarchy will support flexible and adaptable teams. While you won't have the certainty of teams performing given tasks, you'll be able to cut delays, bottlenecks, and absence of creativity inherent in having teams perform assigned and approved work. Guiding teams through vision and purpose rather than a task list will help them take ownership of products and results. In this sort of flat hierarchy, your leadership takes on a role of monitoring obstacles and results, where experts are given the freedom to achieve goals in a way that makes sense to them.

Not every organization can benefit from a flat hierarchy but there are many advantages for software companies including team and individual empowerment. Adapting from a traditional "layered" hierarchy with leadership and management to a flat one without traditional management can be a huge challenge as well.

Many individuals want and need leadership to succeed and getting teams to adapt may result in initial failure. However, once they do adapt, teams operating under their own authority with only goals and objectives to guide them, rather than management and its implied bottlenecks and tasks, will enable them to deliver above your expectations so that your organization grows more quickly.

---

11 https://www.zapposinsights.com/about/holacracy
12 https://www.bloomberg.com/gadfly/articles/2017-04-17/amazon-s-learn-burn-churn-method-puts-zappos-at-risk

Strategy Tip: Flatter organizations with distributed authority stimulate more ownership and engagement.

Team Purpose – the why

Sometimes teams naturally form to fill a need or purpose within the organization. This is often the case in startups, where individuals work together because they either work on the same things, perform the same roles, or otherwise interact with each other frequently. As organizations begin to scale, new teams form in response to a natural need or purpose while others are specifically designed and created by you. In either case, it's important for both you and the team to understand why it exists. To understand this, you have to understand team purpose.

The process of defining team purpose functions in the same way as defining organizational purpose. It gives the team a long-term goal, a greater reason to exist, and an end-goal with which to frame all team activities. A clear purpose helps to define the activities, scope, and responsibilities of the team so that individuals understand not just what the team is doing but where it comes from and why. This is a starting point for everything the team will ever do. It will be important now and as the team scales and grows.

A team purpose could be: "Enable mobile access to customer segment X", or "Deliver self-service customer support for end-users".

This kind of defined purpose help teams to work autonomously and to make their own decisions or take their own direction within the boundaries of the purpose. They are the experts and they know best how to reach goals within any given circumstances. Teams with their own purpose can and will create their own backlogs, generate their own ideas on how to work towards their purpose, and will take the approach that makes the most sense in any given circumstance.

Understanding the purpose of the team also helps teams to better align with the overall company vision and strategy. A purpose clarifies how individual teams contribute to the bigger picture.

While team purpose can be valuable, it isn't always possible to give a single team one purpose. Smaller organizations with fewer teams often require teams to be multidisciplinary and multi-focus. As a result, decision-making and prioritization is harder. However, you can minimize issues by setting a focus with company goals and defining a single purpose for the team based on a relevant goal within a certain timeframe.

It's crucial that any team purpose you define aligns with your organization's strategy and visions as well as other teams. If two teams have conflicting purposes, it will generate internal competition which will be unhealthy for your organization.

Team purpose can help you define team scope, team boundaries, and its products or services. It will help the team to align everything to a single goal, so they know what they are working towards and why.

Strategy Tip: Make sure teams have a clear purpose and manage goals so that each team only has one purpose at any given time.

Team scopes – the what

Setting clear boundaries gives your teams room to operate and act autonomously in ways that benefit the organization. Any team needs a clear scope with which to define their responsibilities, the boundaries of what they are working on, and what is 'their' work or focus. You wouldn't want teams to work on the same things or for work to go unattended, but if you don't define scopes, this is exactly what will happen.

A clear scope defines the boundaries of your team's operation. This includes what the team works on in terms of products, services, or features. For example, a mobile app, reporting module, customer conversion, front-end for mobile apps, complaint process, or exit-customer process. This is different from a team's purpose, because Team Purpose explains why the team exists and Team Scope explains what the team works on to achieve that purpose.

You can make a team's scope as broad or as granular as necessary, but it should be able to keep the team on track with work that directly contributes to the organization's goals without competing with another teams' work.

How you do so depends on how you identify scopes within your organization. For example, one approach is to reduce everything to an internal product, which we discussed earlier. A product can be seen as any asset you deliver to users, internally or externally, providing it has stakeholders and can be improved and maintained. This broad definition is easily applicable to anything from a software module to an entire software product, a process, or even a service.

Teams can then have one or more products in their scope, which they can work to develop, improve, and maintain. Any products inside a team's scope should align with their purpose.

It's also important for teams to take ownership of their products. They should have full product ownership, where they can move beyond maintenance to continuously improve the products in their scope.

What does that look like in action? At Nmbrs, we made feature teams responsible for the mobile app, API, Payroll Engine, HR features, and Finance features. Each module was a full vertical slice of the platform, functionally independent of the other. Each team could own their product end-to-end, allowing them to develop and maintain for end-users with minimum dependencies.

Other teams cannot own the full scope of a product. For example, our UX team is responsible for delivering a great user experience. Their scope includes all Front-end UI on the platform, which does overlap with other teams. This is an unavoidable case of scope overlapping because of team purpose.

My advice is that you aim to give teams full ownership of the scope of their product where possible but be comfortable creating overlaps when it makes sense.

Strategy Tip: Set team scopes to create clear team boundaries, without gaps and with minimum overlaps.

Team size

Teams size affects your organization's stability, speed, and mobility. While teams naturally form around processes or internal/external products in a start-up, you will have to specifically design teams and their size to meet organizational needs as you begin to grow.

It's easy to throw more people at a problem to increase development velocity, but this tactic often doesn't work. Very large teams can be cumbersome, difficult to manage, and difficult to coordinate. A larger number of people can result in confusion and an inability to communicate well in meetings. Smaller teams are therefore important for efficiency, but if your teams are too small, they won't be able to own complete processes.

Amazon's "Two Pizza Rule" is a common recommendation. This rule suggests that if a team or meeting can't be fed by two pizzas[13], or about a maximum of 8 people, it's too large. However, your optimal team size will depend on value, the team's purpose, and the type of team.

For example, a large team is difficult to form and to push to perform in the beginning. Once established, large teams are more stable because they always have individuals to perform needed work, even if someone calls in sick or goes on vacation. More members prevent gaps but also require more coordination and planning. Smaller teams are naturally faster and more Agile, because they're relying on fewer people, can make decisions more quickly, and have less input.

While many people try to start all teams out small and grow them as need arises, some teams simply need to be larger. Why? Introducing change is harder as the team grows, so a team that is adapted to moving quickly may perform poorly once you increase size and force it to move more slowly. Therefore, I think it's better to develop larger teams (up to 8 people) for activities which require stability, such as maintenance or feature development.

New and experimental teams benefit from 3-4 members, because it will be easier to introduce new methodology or follow through on research and development.

Strategy Tip: Develop small teams for innovation and disruption and bigger teams for stability.

---

13 https://www.inc.com/business-insider/jeff-bezos-productivity-tip-two-pizza-rule.html

Planning for team growth

It's natural that teams will grow as part of organizational expansion. It's important to stay on top of changes and additions as this happens, so that you do not end up with too-large teams that well-exceed their scope. This is important because team size impacts productivity, agility, and speed. A team that is too large for its purpose will not be adding value in other places, which will increase overhead.

Splitting teams, building new teams, and expanding smaller teams to meet changing needs inside the organization are necessary parts of team design. Here, it's important to consider the needs you are trying to fill, the scope and role of the team, and why you need a new team. For example, it's often the case that you simply need more developers working on products and features or more customer service agents. It would make sense to simply expand your existing teams, except that doing so could make them too big and difficult to manage.

There are two strategies which you can use to grow teams in a way that best suits the work and your goal – starting a new team and splitting a team. Each of these strategies can be valuable, but typically in different situations.

Starting a new team is the best response when you have a new need, a new feature, or changing requirements inside your organization. If you're pushing massive change, creating new teams can help you to define a set purpose and goal which won't be impeded by previous goals, purposes, or ways of working. Here, you must set a team purpose, create a scope, define roles, and then launch the team by moving people into it. You should not normally onboard new people into a new team, because the large number of unknown variables will cause confusion and problems. With no history of what to do or how to do it, newcomers will likely be confused and unsure of their responsibilities or expected output.

Splitting teams is ideal when you need more people to fill the same responsibilities in either the same or a different scope. For example, if you need more developers to work on the same feature, but the team is becoming too large, you could split a team and have two teams working on the same feature. If you wanted to start working on a new product and feature which required the same responsibilities and skills, you could split a team to draw on the experience and skill of the existing one, without changing its scope. Here, you can grow or over expand the team and then split it, like dividing a cell. You can also split the teams first and then add on new people. Choosing an option should depend on the work and the existing team size.

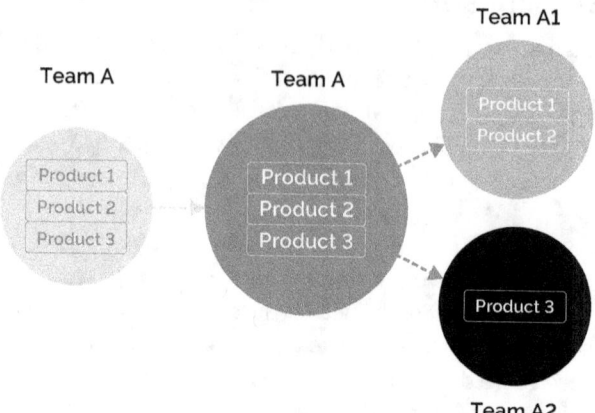

Fig.27. Growing and splitting teams

For example, if you have a 6-person team, you would likely be better off expanding the team to 8 and then splitting it. New people can on board into a bigger group and into an established team, with the benefit of learning how to work with the team before you split into more modules. When you do split, everyone onboard will have experience and you will have less unpredictability in the new team. On the other hand, if your current team is too large for its purpose, you could choose to split a team of 6 into two smaller teams, which can each focus on a different purpose.

In some cases, it's also a good idea to over expand a team. When you need at least one of a specific role (such as a developer), it doesn't make sense to share the role between teams because you will likely add too much overhead for that person in the form of double meetings, a split focus, and conflicting responsibilities. Instead, you should hire on a second person in that role before splitting the team, allow them to gain experience in the team, and then divide the teams into different modules.

Small, specialty teams can seem like a good idea in terms of efficiency and ability to completely dedicate their efforts to a single product. However, this can backfire in the long-term. The more specialty teams you have, the more likely their work will eventually bottleneck while waiting on another team to update code that touches their features. It's also likely to increase overhead if any of those teams rely on a small amount of specialist work.

My recommendation is that you create small, specialty teams – each with their own module – but organize those teams into larger clusters for scaling. As we discussed earlier, Spotify achieves this with "Tribes".

The idea here is that a cluster or "Tribe" of teams work together. They are free to update each other's code when necessary and can make small changes to avoid dependencies. Then, when a specialty team updates their module, which interacts with another team's module, the team making the changes can go in and handle the updates. This will greatly reduce dependencies, enabling you to scale more easily without bottlenecking.

Tribes also allow you to link specific team goals to company objectives, so that a group of teams can collectively work towards a single goal. This allows you to simplify the process of goal setting, which also reduces goal-based dependencies. It allows management to allocate capacity and budget per tribe, so they can drive from a higher level. Squads within the tribe then have the freedom and flexibility to organize themselves within that budget, allocating people based on work such as feature-creation or maintenance.

> **Strategy Tip:** Onboard new people onto existing teams and split modules after they have experience.

Developing cross-functional teams

Cross-functional teams are made up of people and skills from different departments, typically designed around specific modules, products, or goals. While cross-functional teams naturally add a great deal in terms of complexity and management problems, they can be extremely valuable because they work quickly and independently.

Here, my recommendation is to develop teams and their skills around the team purpose. Aligning skillsets to team purpose means you can align roles and responsibilities with achieving goals. This allows you to create cross-functional teams when-needed to achieve team purpose, while allowing you to simplify teams when cross-functionality is not needed.

For example, if you are growth hacking, you need skills from several departments to achieve the team's purpose. Product development teams normally benefit from cross-functionality as well, and I will focus on this team as an example for the rest of this section.

While designing cross-functional teams around their purpose can feel like a nice-to-have, it also prevents some of the issues that crop up with these teams. For example, many cross-functional teams struggle with alignment, leadership, and collaboration. Assembling a team around a single purpose and focusing on alignment from the start will work to circumvent this.

Product development teams are the most crucial point in your development pipeline. It is their work, creativity, and ingenuity resulting in products and therefore in delivering value to the customer. Creating a product development teams structure around that value, rather than around simply creating features and developing new products is a crucial element in operations. Developing a cross-functional team (a squad) is the ideal solution for product development, because it enables you to better utilize skills across departments rather than forcing teams to transfer work back and forth.

Your goal should be to develop a squad that integrates the skills capable of driving the most value from development. The squad must be more than the sum of all its parts. You must balance roles across squads, combine roles in beneficial ways, and design squads around the product being developed.

One valuable consideration is integrating support consultants. The support consultant can bring the user's voice and perspective to development, because they are exposed to support incidents and how people use their modules and applications. This will help the product development squad to better connect to users and what they actually need, while creating a channel for customer support to better understand how the product works.

I also recommend integrating Quality Assurance (QA) into cross-functional product dev squads. In traditional waterfall, the developer will produce something and the QA will evaluate and test it. In Agile squads, QA should work with the squad to contribute towards solutions within their own specializations. Here, QA should already understand use cases for that part of the feature and can begin implementing testing and validation before the developers have something to test. This sort of testing will always fail at first, but as the work progresses, it will pass.

This process allows QA to directly contribute to solutions, where people who understand functional needs and the end-user are part of the coding process. It's also a huge step up from the old waterfall processes of having a functional analyst draw something, having a developer write code, and a QA test it, which can often result in products which are out of touch with the user.

Involving QA from an early stage will allow them to offer advice that brings the user experience into perspective from the start. It's also important that QA understands how the user activates or implements products, how they will use it, and where it will be utilized. This ties into the first point of integrating a support consultant into product dev, as support is the most logical place to source this information.

Importantly, integrating QA in this way means that there is no "tester". Instead, the QA should take on the role of quality engineer for automated functional and nonfunctional validation. Because everyone is involved with the product from the start, everyone has the knowledge to test whether the product is functioning as needed. Creating processes where everyone is expected to test product functionality will push developers into testing front and back end code, so they can connect the product to the value concept and the end user's need.

This is vastly different from Waterfall, where developers often don't even open their product, just look at the code. Agile forces them to be more involved and to understand the actual value of the features.

Designing squads in this way will help you to reduce dependencies, speed up development, and increase the total quality of the product. For example, back-end squads typically depend on QA testing because they lack functional knowledge of features, so QA must validate the impact of their work. Integrating QA as engineers and automating most QA processes avoids this step, because back-end squads can run automated tests and build products based on input from QA from the start.

While Product Dev is one of the most important types of cross-functional teams you will have, you will need many others, and team-members will vary a great deal. Here, I believe that two roles should always be present. Scrum Master to optimize processes and reduce impediments, and product owners.

Cross-functional teams allow you to take a different approach to assigning Scrum Masters. Many organizations assign a scrum master per squad. Here, individuals don't have to learn internal organization, because the Scrum Master is handling it. It's much more efficient to design multiple squads with a single, external scrum master.

I believe that a scrum master should operate with the goal of making himself obsolete. However, some complex squads will always require their own scrum master.

Any squad, but especially Product Dev, should also have an assigned Product Owner (PO). The PO is responsible for external communication, so that if you want to know what the squad is working on or where they are in development, you go to the PO.

Designing cross-functional squads around their purpose creates teams capable of aligning on a goal and achieving that goal themselves. With no dependencies or bottlenecks, these teams can rapidly develop, innovate, and grow modules and products. However, companies, markets, and products change. When the team purpose changes, it's important that the team be dynamic enough to change roles and skills, so that it continues to function around its purpose.

Our onboarding squad at Nmbrs faced relevant challenges. They were responsible for onboarding new customers into the product. This meant developing tooling into the product, so customers could start on their own, without help from customer support. To make this happen, we added an onboarding specialist, responsible for monitoring new-user activities to the team. Adding this role on top of Product Owner, Developer, and QA engineer meant we had someone who could offer guidance to the new user while bringing input to product development. Using that input, Product Dev was able to make the process easier over time.

Strategy Tip: Design teams around their purpose, pulling from cross-functional disciplines when necessary.

Working with Specialist teams

Specialist teams are designed to either contribute to a single goal or to one part of a process. These teams are common in old waterfall-style organizations and can be both advantageous and disadvantageous. On the one hand, they isolate specialist experts such as UI/UX designers, developers or support consultants into one "room", effectively limiting their ability to own processes as part of smaller teams. On the other, they bring those experts together from across your organization so they can better share ideas, collaborate, and work on organization-wide solutions.

The best way to design teams for startups is to create broad scopes, so teams can do more. This means you should begin with generalist teams and add on supportive specialist teams as your organization grows. Later, this will evolve to include experts on each team when you can afford them.

Why? Most organizations don't benefit from teams that group all specialisms or functions. Your UI/UX designers will add more value when spread across the organization and integrated into other teams. In some cases, it's not necessary to have a full team dedicated to a process. In others, the practice of creating Agile teams, capable of complete product ownership, necessitates specialist distribution across teams. For example, if your team were responsible for delivering features for a platform with defined design guidelines, the team would need a full-time UX/UI designer.

Splitting specialists up across teams makes sense in many ways. However, it's still important for specialists to be able to communicate with each other across the organization, to quality check and audit each other's work, and to share ideas and solutions so that silos don't develop.

Specialist Teams

Dedicated teams are a common early strategy for integrating specialists. Here, you can choose to create a dedicated specialist team to support other teams in their work. For example, a UX team designing UX and UI patterns and guidelines for a certain product. Their work could be utilized by many product teams, benefiting the entire organization. This is one of the best approaches for smaller organizations who cannot afford to integrate a specialist per team, or when it's extremely easy to share work across the organization. This approach is also valuable when you have a large amount of work to be completed by specialists.

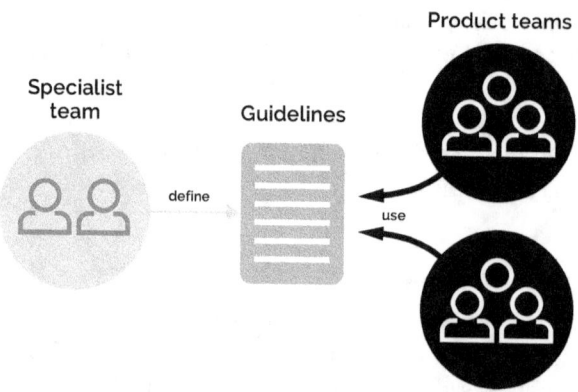

Fig.28. Working with specialist teams - deliver guidelines

Specialist Chapters

Another common strategy is to create a "Chapter", from the Spotify model, where specialists are allocated across teams. This allows specialists to add value to teams so they can be autonomous, capable of complete product ownership. If you have 6 product squads and you assign two UX designers to each, you can define a chapter to connect all UX designers, no matter the squad.

Why? While it's always ideal to have a specialist in each team, you cannot logically drive enough value in the early stages of scale-up. You likely won't even have the support structure or workload to make those experts valuable. Specialist who operate across larger Chapters solve this problem by allowing you earlier access to specialist skills and functions. However, this can backfire because specialty tribes have to involve themselves in every team and police each for quality/security/user experience/performance, etc., which can greatly increase overhead.

Product teams

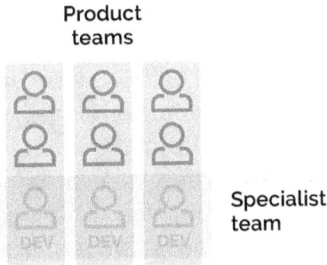

Specialist team

Fig.29. Connecting specialists from different teams

Linking specialists across squads will help avoid silos. It will enable knowledge to be shared across the organization so that guidelines, approaches, and processes can be defined and reused by everyone. It also helps you to create internal guidelines for quality management, processes, and guidelines across specialties. You can use peer review, because each person works inside the same specialty and will easily understand what other members are doing inside the role.

It's not a good approach to give chapters their own backlogs because this will result in the individuals having a double backlog. Instead backlog should go to a squad. Chapters are stakeholders for the squad and contribute to that squad. Having chapters capable of contributing to squads is important because the only time a specialist team should own a full product is when it is entirely in their specialty.

The specialist completes technical work for their squad. However, in their specialist chapter, they should be responsible for defining and managing guidelines, quality metrics, code review, best practices, mentoring, and onboarding on their specialty, where they can truly add value.

Strategy Tip: Link members of a business function or technical specialty into a chapter to ensure they can share ideas across the organization.

Enabling Your Teams

One of the tenants of Agile work practices is that teams become self-sufficient and able to handle their own processes, management, and infrastructure. This allows one team to take total ownership of a single internal product, so they can make decisions without bottlenecks. Complete ownership of an internal product offers a great deal in terms of benefits, but it has drawbacks as well.

Here, a team will have to step outside of their specialties and even relevant skills to create infrastructure and processes for their product. Even people managing technical work such as coding must be able to take on new tasks and responsibilities. This is a natural result of approaching work from the aspect of what you need to deliver.

Very flat organizations require teams to take on more and more responsibilities outside of their primary role. This can branch out into side and support activities which they are not technically skilled in, but which contribute to the team owning the process. For example, it's very common for Agile coaches to step in and ask squads to handle their own recruitment, onboarding, holiday management, process design, development for data dashboards and so on.

It's undeniable that performing this work is essential to enabling teams to succeed. However, performing this work can bog teams down in these activities, taking time away from work that adds value. If squads are doing too much of their own support, they won't be doing enough of what adds value.

An easy analogy for this problem is that of a city. People can go there, trusting that they have access to electric, water, and sewer – simply using the existing system. If something goes wrong, they can call and have it fixed and then go back to being a system user.

This is a stark contrast to some Agile squads, which are asked to manage and maintain their own "utilities". When something goes wrong, they have to diagnose the issue, find a solution, fix it, and make it better. They can't simply rely on the "electric" being there to power their computers for work, they must take an active role in it. Squads can begin to struggle because they aren't experts in hiring, running processes, or onboarding, any more than the average city dweller is a trained electrician. At the same time, you can't afford to assign operational experts to each squad or team, because this would naturally create too much overhead.

The solution is to enable teams to focus on delivering value to the end-user, by reducing their overhead work and keeping responsibilities.

Here, overhead teams play a valuable role. They provide the support and processes the squads need, so squads can operate primarily as users. If a squad is taking too much responsibility for structure, they won't focus on core responsibilities and you will lose value. Overhead teams, like HR, operations, or finance teams, can step in to manage processes, hiring, and onboarding. Ideally, these teams will work with the squads to ensure what is being developed works for the specific squad and is delivered in a way that meets their needs.

While overhead teams are not directly adding value to the customer, they do so in other ways, making it possible for core teams to perform their work. They clear the way for squads to focus on delivering value. Without them, your developers and engineers would lack structure, support, or even finances to focus on their purpose and goals. While squads can take on those responsibilities to a certain extent, you will lose value and productivity if they are spending too much time on structure.

Overhead teams provide value, but it's important to create and achieve a balance. When overhead teams become too large, you may be putting too much focus on developing a beautiful organization, without an end goal or value. On the other hand, when overhead teams are too small, core teams will not have enough support and may underperform.

> **Strategy Tip:** Implement overhead teams to create structure enabling other squads to focus on value-added work.

Component vs Feature Teams

Component teams become relevant as your organization grows. Teams naturally have to create and manage a lot of basic components, platform updates, security layers, logging, internal frameworks, etc. This work is necessary and valuable, but it detracts from end-value to the user – delivering new features. Component teams step in to take over this work, performing in a cross-team role to deliver the supporting structure and platform functionality that enables feature teams to deliver their features.

Where overhead teams handle organizational support such as HR and finance, component squads perform enabling work in a cross-functional role. You can think of this work as platform improvement, where component teams focus on delivering cross-cutting and shared components which feature teams rely on to increase their productivity.

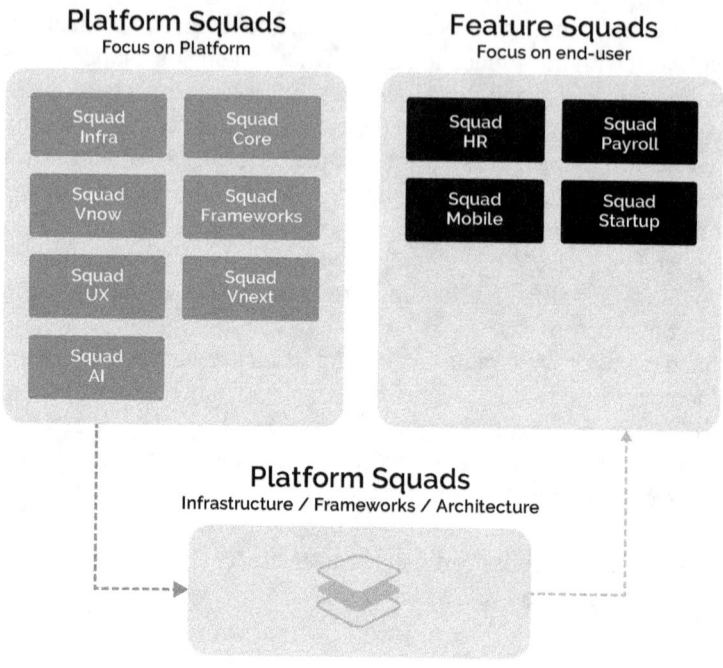

Fig.30. Component teams enabling feature teams

In our metaphor, component teams work to improve the electric grid, increase the efficiency of the system, and otherwise improve your base infrastructure. This applies to everything from improving security across the platform, to boosting performance, to working with code quality, to UX, to infrastructure.

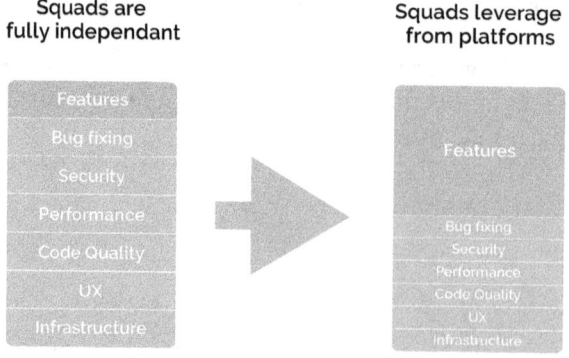

Fig.31. Optimizing squads for feature development

This adds value when you have enough existing feature teams in place that a components team will always have enough work providing basic components.

The benefit of adding component teams is that feature teams can spend more of their time delivering features. However, doing so doesn't change squad responsibilities. Teams should remain responsible for their own infrastructure quality. Component teams simply add support or help for these responsibilities. This allows you to maintain quality of implementation, because teams always have to pay attention to security, infrastructure, and other factors, even when they aren't directly working on those elements.

While both component teams and feature teams are supportive teams, each adds value in different ways. Both make room and offer support for the teams delivering direct value to the customer. Component teams focus on delivering a solid and scalable platform, while feature teams focus on delivering features for end-users.

Strategy Tip: Develop component teams to help product teams deliver direct value to end-users.

Maintenance and Innovation Teams

Both innovation and maintenance are crucial to continued performance and business success. These two factors often contradict each other and may impinge on each other's ability to operate. It's difficult to handle maintenance and incremental improvements while innovating. Innovation naturally brings disruption and this doesn't contribute to stability. Maintenance teams exist to keep your core stable and reliable. They will naturally clash with teams dedicated to innovation, who exist to discover new approaches and create bigger steps forward to better and more efficient operations.

At Nmbrs, we design teams based on their operational focus, which I call "Oil Tanker" and "Speedboat" teams. Oil Tanker teams handle core processes such as maintaining a feature or function and developing new features for it. These functions and processes must remain stable in order to continue delivering value to the customer. They have to move slowly and reliably, like an oil tanker. Oil tanker teams are large, stable, and able to consistently perform work in a way that is reliable, but which doesn't necessarily support rapid growth.

In other cases, the goal is to grow quickly, explore, discover new things, and innovate. Here, we apply smaller, more flexible "Speedboat" teams, capable of making agile leaps, taking risks, and truly innovating in big ways. They're not as stable, but they're fast and agile, like a speedboat, which enables them to move quickly and in ways the oil tanker team wouldn't be able to.

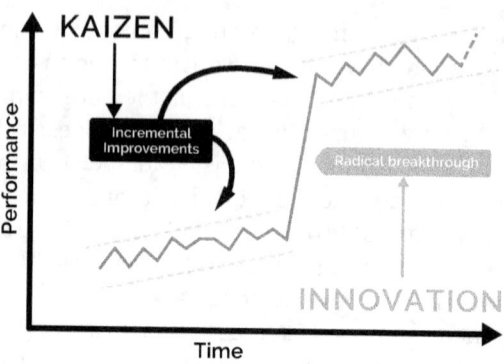

Fig.32. Kaizen vs Innovation

While there are many factors to keep in mind – for example, it's important not to introduce innovation for the sake of innovation – there are compromises that make them worthwhile. Small and lightweight innovation teams – capable of implementing innovation in short-term, short-investment-cycles for rapid adoption, testing, and implementation – will reduce risks while creating more opportunities for improvement and even creative and preventive maintenance.

At the same time, it's important to manage what each type of team is doing. It's tempting to have Oil Tanker teams just doing maintenance and Speedboat teams who just handle innovation. This might seem like it will solve problems, but it will likely contribute to dissatisfaction in your teams, because a large portion of employees would just be fixing other people's work. In addition, creating maintenance-only teams removes innovators from accountability – reducing their drive for quality assurance – because they lose touch with what they are doing and why it has to be stable.

Instead, both teams should handle maintenance and development in different ways. Developing new features is part of software maintenance because you're not maintaining a permanent feature, you're maintaining customer satisfaction. Maintenance teams are not just fixing bugs, they also contribute to developing new features.

There's always a point where innovation and maintenance overlap. In some cases, one team will need help from another. An Oil-tanker team might simply be patching a leak and not solving a problem. They don't have the time, focus, or resources to delve into one problem because they have too many modules to maintain. Bringing in a speedboat team could add the innovation and flexibility to solve the root-cause. On the other hand, there are many instances when fast-moving speedboat teams won't be the best choice. For example, if you want to maintain quality, performance, and stability, disruptive innovation is not the best approach.

What truly differentiates these teams from each other? Oil Tanker teams typically take small, incremental steps, focusing on immediate problems, updates, and features. Speedboat teams tend to take much larger steps, implementing innovation and large change as part of their work-process.

It's important that your teams fit their roles, because each maintenance and innovation require a vastly different mindset. Some people will naturally fit into day-to-day development, creating new features for now, and working to improve customer satisfaction now. Others will be better suited to developing new solutions that dramatically disrupt and innovate on existing ones.

Your role will be to organize a structure coordinating the efforts of Oil Tanker teams to keep operations running smoothly alongside Speedboat teams' efforts disrupt and improve in bigger steps.

> **Strategy Tip:** Balance innovation and maintenance teams so that neither gets in the other's way.

The Leadership Team

Your management or leadership team will have a great impact on the rest of the organization no matter how flat your organizational hierarchy. While the roles and responsibilities of this team are diverse and will change from organization to organization, your management team will influence the company through inspiration and leadership style, direct feedback and orders, and behavior.

Over time, this influence will greatly impact the culture, focus, and behavior of the organization. Is the CEO always late for meetings? Other leaders will find it acceptable and replicate the behavior, which will then trickle down to the entire organization. Is the leadership team unable to share clear decisions and structure? The organization will be confused because people will find it difficult to recognize and take their responsibilities or to be productive.

This is very relevant for the Tech COO. The full operational model must be implemented across the entire organization, from top management to the work floor. Everyone has their own role in such an operational model; therefore, it needs to be understood and implemented by everyone. Your management team is key to pushing that operational model down or across, depending on your hierarchy.

It's important to describe the role of your executive team in your operational model and then to push for adoption. Executive teams are important stakeholders for delivering quality and productivity, and every member should have a clear understanding of their role and responsibilities in your value stream, so that they can contribute and work for adoption across the organization.

An Agile work environment demands a relatively flat hierarchy, which changes how management teams interact with other teams. In an unhealthy organization, management can be extremely disruptive because people may feel that questions must be handled at the expense of existing goals and tasks. In a healthier organization, the management team works with everyone else and doesn't expect to be followed. Management teams should contribute at a strategic level, including budgeting, higher-level work structure, goal setting, and theme focusing. Everyone should know their responsibilities and management should know theirs, so that management does not disrupt work.

At the same time, management should be an active stakeholder, directly contributing to goals and output, and not just a spectator managing output.

Strategy Tip: Involve the executive team in your operational strategy.

Roles and responsibilities

Once you've designed teams to run operations, it's important to define how work will be completed. When you have a product development team, the individuals must collectively have the necessary skills for the team to deliver their purpose. In this case, the team needs to be able to develop software features for a product. Defining roles and responsibilities is critical for HR and management, who utilize responsibilities for hiring, team management, and so on.

It's important to understand how a responsibility fits into the value stream. Here, you can picture the value stream as a "factory" with production steps. If you add a role, you have to know how and where it contributes. Responsibilities define the how and where.

Teams that don't understand their responsibilities will not take ownership of modules and decisions. This is very often a problem in fast-growing companies, because teams grow quickly, operational strategies change, and roles are often not re-designed to fit changing purposes. The result can be underperforming and sometimes superfluous teams.

It's often unclear if defining roles and responsibilities should be an operational task or an HR one. This does depend on the organization's strategy and vision, as well as the scope of each department. You may consider collaboration. No matter what you choose, it's crucial that you define responsibilities and quickly.

Your organization must describe the roles needed to fulfill a team's purpose to ensure a good operational fit. To do so, you have to take operational strategies into account – as well as the team's goal or purpose – when defining the role.

Here, you can easily integrate standards and competency frameworks. These predefined frameworks list skills, behaviors, and competencies which contribute to success inside of roles such as developer, support consultant, QA engineer, etc. However, this will largely be an HR rather than operations responsibility.

Defining responsibilities as part of roles will make it easier for you to design processes which align with and are framed within such responsibilities. If you know your team's responsibilities, you know how they work and why, so you can create better processes. This is crucial, because processes and roles must be precisely aligned. Process implementation may change, especially for optimization purposes, but it normally stays within the scope of existing responsibilities. Deeper process changes will require you to review and possibly change responsibilities.

Assigning individual responsibilities inside the value stream also makes it easier to monitor performance, motivate team members, and enable individuals to take ownership of work.

Strategy Tip: Clear roles and responsibilities will help motivation and engagement.

Responsibility vs accountability

Agile teams can move quickly and without the impingement of hierarchy. Agile frameworks give teams more ownership of products by removing leadership positions and allocating responsibility across the technical persons capable of developing those products. While this can speed up production in many ways, it's important to maintain a standard of accountability by ensuring that individuals know what they are responsible and accountable for.

Using a clear model to define not only responsibilities (who should do what), but also accountability (whose fault is it?) makes it possible to measure and define quality, to hold teams accountable for quality, and to correct issues when they appear. When something goes wrong or must be addressed, you need to know who to go to. Teams also need someone who has final say on decisions, so that someone is always accountable.

Many Agile teams struggle with accountability because the methodology works to distribute accountability across the team. It's an interesting paradox because the logical solution is holding the entire team accountable. Unfortunately, this often doesn't work in practice. A person accountable for quality or output will monitor and report on certain tasks, activities, and processes. When everyone on the team thinks everyone else is accountable, each person might think everyone else is paying attention, often resulting in no one paying attention.

Assigning one accountable person will help you circumvent this issue. The Product Owner is one role which is most commonly held accountable. The Product Owner is more external to the team but is also in control of backlog, priorities, and decision-making, which can influence team performance.

The RACI model is a common framework used to define who is in charge of a product and its quality inside any given squad or team. RACI or "Responsible", "Accountable", "Consulted", and "Informed" fits easily into process design and management, giving you a simple tool to let people know their role in the product, tool, or its development.

Here, Responsible means that you will carry out the decision and do the work to make it happen. Accountable means that if the decision goes wrong, it's your fault. You get to say yes or no to decisions for this reason. Consulted means that the Accountable person asked for advice and considered your input. Informed means that you are aware of the decision but are not consulted.

Whichever model you choose to integrate, it's important that you have defined decision-makers for each product or solution, so that you know who to go to when something does go wrong.

Assigning a single person who is accountable for a decision also works to prevent instances where decisions aren't made, because no one is willing to take on responsibility. No one wants to be responsible when something goes wrong, so Agile teams often stall because decisions are being pushed off, with the assumption that someone else is making them and taking on accountability.

Assign a single accountable decision-maker to ensure decisions are being made and that when something goes wrong, someone is accountable.

> **Strategy Tip:** Assign accountability to only one person. If more than one person is accountable, then no one is accountable.

Multiple hats

As a startup or small organization, it is unlikely you will be able to hire for every possible role you can fill. Many startups ask team members to fill several roles, wearing multiple "hats" to accomplish what would, in a larger organization, be the work of several people. Because the nature of a small company means the volume of responsibility across each role is lower, assigning one person to multiple roles works – at first.

While the volume of work is low, you shouldn't have to hire for every needed role. However, even if you don't have the capacity to hire for every role, it's important to consider you will eventually have to.

If you continue to grow and want operations to run, roles will split, and more employees will be brought on. You have to design roles for the future, when you have the budget and workload to allocate responsibilities in a way that aligns with how you want operations to run. If you do so, you can expand more easily when operations begin to grow.

This will be massively cost and timesaving over designing roles and responsibilities around the people you have now and reorganizing operations later. The smaller your operation, the easier it is to structure organization. If you have to design your roles as you need them, you won't have the time and resources to do so without significantly slowing or diverting from other operations.

How can you make this work before actually having a larger organization?

Create the roles you will need, allocate several roles to one person, hire on new people to fill roles as they grow, and ensure each person understand the several "hats" they have, so they can act accordingly. This will give you a better scope to understand organizational growth and where to hire. It also helps you understand where and how to move existing people as their roles change or the scope of their role narrows.

Having a role framework in place helps communicate what a role is, which roles an individual is filling, and how individual responsibilities change when they are moved out of multiple roles.

Organizing your operations in this way also means hiring the right people. For example, it can be tempting to hire a specialist who is very good at something such as UX design to solve internal problems or simply to make things better. Unfortunately, without the infrastructure and large teams in place to support a specialist, they will have to expand into generalist work and will likely be dissatisfied with work in your organization. My advice is to start out hiring generalists rather than specialists to fill roles, and then begin to add on specialists as you split roles and build teams when demand grows.

Strategy Tip: Allocate responsibilities to roles instead of people, so that you can split roles later, without redesigning them.

This ties into the concept that it's important not to hire someone until your organization can benefit from them. Hiring at the right time is just-as-if-not-more-important than hiring the right people. Defining the right time and communicating those needs to HR will help you to prevent high churn and high overhead by ensuring people are brought on when they can add value but also when you can provide the structure and support enabling them to do so.

Assigning multiple roles to individual people allows you to assess needs based on the individual's ability to manage work. If one person handling multiple roles can tackle a job, it's likely the wrong time to hire. You may benefit from hiring a specialist, but they might not yet be able to make the most of their skills inside your organization. If you hire someone who is very good at what they do but you don't have the structure to enable them to focus on it, they will eventually become unhappy and will leave.

At Nmbrs, we wanted to improve our user experience and interface. We hired a UX designer to begin this process. While his work was very good, we didn't have the structure to support him, he was a solo expert in a team of generalists and it eventually didn't work out. If we'd waited until we were able to afford and build a dedicated UX team to support him, he could have brought much more value to our organization.

Specialists want to do specialist work not generalist work. They often aren't comfortable taking on multiple roles or doing general work. In a small company, you also often cannot afford the overhead of a person who only contributes to one thing, even if that thing is important. Instead, it's better to wait and set up the structure supporting what that person can do before you bring on a true specialist.

This also applies to actual HR teams and finance, because these teams and hires contribute a great deal to overhead. Many people create an HR team immediately, even for a few employees, but this creates unnecessary overhead. Instead, I would recommend creating the structure for HR, distributing HR and finance responsibilities to founders, and only hiring on HR specialists when management becomes a problem.

Strategy Tip: Ensure that the organization is ready before you hire for or implement specialist roles.

Team dynamics

Structuring teams inside the organization is one step to ensuring productivity and creativity. You must also work to structure individuals inside the team. Each person must complement other members of their squad so that they can work together in a fast, efficient, and creative way.

While only direct leadership will have true insight into how individuals work together, building structure and processes for fitting people together will help your organization be more productive in the long-term.

Several models or frameworks exist to help you achieve this. One of those is Belbin Team Roles, where work personalities are mapped and assigned to individuals who fit them. Here, many individuals fit into several roles, allowing teams of 3-5 people to easily fill all the personality traits that make up a good and productive team. The Myers-Briggs Type Indicator (MBTI) is another commonly used framework, which uses four dichotomies or eight personality factors to assess how individuals interact and handle situations. These personality factors are then used to divide individuals into 16 personality types, indicating how and why they work and work together.

Another consideration is that teams should bring some balance, variety of thought, and cognitive diversity so that individuals can challenge each other to create new and better things.

One large problem with traditional teams is that many people work in silos, are often grouped together based on what they do, and therefore are not exposed to new ways of thinking or doing. They risk becoming static and relying on tradition and formal process rather than Agile thinking.

Introducing and managing cognitive diversity as part of team structure is one way to combat this problem. Cognitive diversity includes any difference in perspective, such as age, information processing styles, level of education, nationality, and even work roles. Knowledge processing, or how individuals prefer to use existing knowledge versus creating new knowledge in new situations is one big factor. Perspective, where individuals either use their own expertise or leverage that of others is another.

Introducing cognitive diversity into teams means creating a deep understanding of the individuals on your team and of individuals during the hiring process. This will require a strong cooperation with HR and team leads who can directly contribute their knowledge of behavior and responses as part of performance or competency management. It's also important to ensure that team members get along and can work together. Diversity for the sake of diversity may create friction and confusion as workstyles or thinking and processing clash.

Strategy Tip: Make sure team members are compatible and there is a balance in cognitive diversity.

Assigning people and teams

Any team will be made up of a diverse range of people from different cultures and backgrounds. Many will display different personalities, different types of intelligences, and different thought patterns. We discussed this earlier in team dynamics, but it's important to keep in mind when creating strategies, assigning work, and even guiding people in how they work.

Here, you should work with HR and a set competency or behavioral model to define how people work, how it affects your teams and strategy, and what skills and strong points team members display. You can then leverage these skills to empower teams to work in a way that is most productive. For example, many people are very good at improving existing work but do not excel at creating something new, others are the opposite. Putting both on the same team and expecting them both to do the same thing would be disadvantageous for both.

Understanding how and why individuals work can allow you to make better choices when hiring, assigning work, and when letting teams choose how to work. If you know which competency and behavioral factors contribute to success inside a role, you can connect with HR and hire someone who matches that profile.

Similarly, it's important to pay attention to culture and core values. Most organizations eventually have a set culture, where most or all the people onboard believe in the same things, hold the same core values, and have a similar moral outlook. It's important for HR to recognize this and hire accordingly, because people brought into an environment that doesn't match their culture won't be happy and they will (likely) eventually leave.

Aligning people at a fundamental level (such as your cultural values) will enable you to keep teams, work, and people on track. For example, performance in daily tasks and processes directly links to work ethic and time management. People must be willing to make decisions and act in accordance to your core values and culture, so that their intentions always match up with company goals. Once people approach situations with the same intention as you, they may only need (learnable) skills and experience to grow to the next level. On the other hand, values and mindset are not easily learned and someone with different intentions than you might not be able to adapt.

The top players in any organization are those who align with core values and culture but who also have a margin for growth so that they can expand expertise and skills with the same intention as the organization's purpose.

> **Strategy Tip:** Design teams around how people work. Make sure that individuals work with the same intention as the team or organization.

Assigning work is another important part of operations, where you allocate your people to the teams who most need them. Because people are one of the most important resources in any company, especially a SaaS company, this process will impact your ability to function as an organization. Assigning people and teams often overlaps considerably with HR and you should work with your HR team to do so.

Balancing work assignments between individual preference and where individuals add the most value is possibly one of the most crucial aspects. Having the right people in the right place makes a huge difference to teams and your operation. At the same time, everyone wants to work on the most fun or "cool" thing, and everyone has their own personal preference and profile. An individual may add more value in an area that isn't their preferred role or team. Trying to match people to where they add the most value while balancing it with where they will be most happy can be tricky and it won't always line up. You may have to put people in places that benefit the organization, but which may not be where they want to be, which will eventually result in turnover.

Reducing this risk means talent scouting, using tools like competency and frameworks to determine how and where people want to work, and hiring and assigning people based on what you need for the role rather than what they can do.

This will also require balancing obligations with HR. In some organizations, operations assigns people to teams, in others, HR asks individuals what they prefer. Allowing people to choose always means creating happy teams, but at the risk they might not contribute to their fullest. If operations handles assignment, you can technically apply skills where you most need them, but may lose team satisfaction and therefore productivity.

Finally, some placements never work for reasons that are outside of your control. An individual may not get along with their colleagues and may quickly become demotivated. They may realize the role wasn't what they thought and become demotivated. Matching people to roles and teams where they can be most productive is an ongoing and continuous process.

You should always assign individuals based on a combination of risk factors and qualifications. Putting new people on new projects never works because there are too many variables and unknowns. Instead, you should place new people in very solid teams while you determine what they can do and how they can do it before moving them onto a new team or project. This will help to ensure that your risks relate to the project and not to the team when starting something new.

> **Strategy Tip:** Don't assign new people in the company to new projects, make sure only one is an unclear variable.

Recruitment and Operations

Managing people and structure is very much an HR responsibility but it heavily overlaps with operations. Here, the goal should be to create processes that naturally select and match people to an environment and team. Creating these processes will be important for ensuring long-term productivity of your most valuable resource (people).

Different people excel in different types of environments and will appreciate more or less structure. Some people only perform in well-structured and defined environments, others excel without it and at creating it. Your scaling organization will have space for both, but you need balance. The risk of scaling is that it will create a lack of structure. This controlled chaos can make room for innovation and disruption. However, too much chaos will be disruptive and confusing in a bad way, especially for individuals who need structure.

At the same time, too much structure hampers progress. If people are bound by strict processes and structure, they can only maintain processes and won't have the freedom to create new and better ones. Very creative persons may chafe at too much structure and will leave for a more open or creative environment.

Your goal should be to define what type of structure you need and are setting up so that you can match people to your structure. Working with HR to define what you are looking for will enable you to match individuals to your organization, so that you can work well and in the same way.

Some people need a great deal of structure with a manager and guidance. They will not likely excel in a flat hierarchy and might not know what to do with the freedom and autonomy offered by Agile work practices. Hiring someone who already works in an Agile way will prevent frustration and lack of productivity by ensuring that the person you're hiring actively wants to work autonomously.

Strategy Tip: Hire people who can work in a way that meets your current and future organizational needs.

The concept of hiring people based on competencies and soft skills such as values is an important one and it should be integrated into work processes. Why? While you can always train for hard skills such as a development, you cannot change why people work. If someone is working because they believe in what you are doing and want to see it succeed and want to be proud of what they are doing, their work will naturally be different than someone who is working but not particularly motivated or inspired to do so.

For example, many people move into organizations because they want to escape their current job. They have potential but their only real driver is that they want to leave their current job. They see your role as an opportunity to leave, not an opportunity to grow or contribute. If you feel that this is the case, it's never a good reason to hire someone.

Creating a clear and transparent hiring process, where potential recruits are aware of the organization's purpose and potential journey, is one important way to ensure that employee motivation and goals line up with those of the organization. If people can see the organization's journey, they can become interested and excited about it and therefore much more motivated to contribute and join for good reasons, because they want to join the organizational purpose.

At Nmbrs, our best people are those who joined not just to code but because they wanted to be part of the organization. They saw, liked, and believed in what we were doing and wanted to be part of it.

> **Strategy Tip:** Ensure that new hires are aligned with organizational vision and are motivated for the role.

Juniors vs Seniors

HR takes on a large portion of decision-making when hiring for roles. However, operational structure defines who is working in your organization and why. Here, one major point is whether you plan to hire juniors or seniors. There are pros and cons to both.

Juniors are often just starting out, may be recently graduated or even still studying. They're young, typically under 27, and, while they don't have the same wealth of experience, often bring creativity, a knowledge of new technology and work-methods, and a fresh perspective. Junior hires are also readily available, affordable, and easy to quickly hire and train into the employees you need.

Seniors are much more difficult to hire because demand is typically high. With a great deal of working experience, senior hires know the industry, can provide structure and organization, and can bring a wealth of knowledge and experience to solve problems, build creative solutions, and organize teams. This can become problematic if you hire from waterfall rather than Agile organizations, so senior doesn't always mean advantageous in a startup or scale-up organization.

While both seniors and juniors have their advantages and disadvantages, I believe it's important to find a good balance. You ideally want to bring on a team consisting of different experience levels, where juniors can learn from and follow seniors, and seniors can be challenged by the new perspectives and knowledge of juniors. However, this balance will depend on your team, what they are working on, and the level of skill, autonomy, and knowledge required by each team.

Defining a good balance between junior and senior employees means identifying where each contributes to your value stream, how much guidance and autonomy is required for the team, and how much you can invest in training very inexperienced personnel.

A golden rule is to use a ratio of 25% juniors, 50% mid-level, and 25% seniors. This ensures you have enough seasoned people to leverage their experience without creating conflict, plenty of mid-level employees to handle most of the real work, and a younger generation growing and onboarding for the future.

It's also important to consider the individual. I personally prefer a junior with more potential over a senior who may have stagnated and who will refuse to adapt or change. Playing the infinite game means assessing future value and contributions versus ability to contribute now, in which case, hiring a junior with the capacity to grow and contribute in new and exciting ways will always add more value to an organization than a senior employee who can apply value once, and that's it. While this won't apply to every senior employee, it's important to evaluate everyone's potential for bringing long-term value, their ability and willingness to adapt to change, and their value now.

Strategy Tip: Define a balance between junior and senior employees and organize teams accordingly.

Working with external people

External teams and people are increasingly common in organizations of every size, with freelancers, agency hires, and outsourced teams all contributing to even the world's largest organizations. These teams are also increasingly practical and essential for businesses of all sizes, because they contribute value in different ways depending on the organization.

For example, small organizations are often unsure of workload and are therefore unwilling to bring on a full-time hire. An external team or freelancer can give you the freedom to add on work as needed, so you can grow in a scalable way without dramatically increasing overhead.

As you grow, external roles allow you to bring on specialists and experts who won't add value over the long-term, but who will in the short term. You can also hire on more employees quickly and for shorter term projects such as product launches, where you won't need the additional employees for the long term.

While external teams are a risk and a commitment you might not want to take, they do add value and they are contributors just like any other member of your team. The risk here is often that organizations separate external workers from both the organization and the work being completed.

Most teams will work alongside external workers in the same room and using the same desks, but those external workers often bring their own computers, are paid in different ways, may have a different work method, and may even have split work or managers. This isolates them from the team, preventing good collaboration, and sometimes preventing the individual from becoming truly invested in their work.

If you want to work with external people for whatever reason, that person should be fully onboarded and part of the team. Doing any less will net you an employee who is less than invested in your organization or its goals.

My advice here is to incorporate external people into the same communication channels as the rest of the team. This will help everyone to feel as though they are part of the same team.

My company Nmbrs works with an external DBA (Database Administrator) expert. His role is to be available whenever our developers needed him. At one point, I realized he wasn't being utilized, most people weren't going to him when they had database questions or problems, and he wasn't actively participating with most relevant projects. I introduced him into our internal chat system over Slack, which quickly resulted in a reversal of the situation. People interacted with our DBA as though he were working in-office with them, resulting in a great deal more value for everyone involved.

External work is becoming more and more common and it is valuable for most organizations. Working to create an environment in which an external worker is just another contributor so that they can be just as passionate and engaged in teamwork and output as a full-time employee, will benefit your organization.

Strategy Tip: Treat external employees like internal ones.

Office environment and productivity

While more and more employees are working remotely, from home, or in flex situations, most will spend much of their time in your office. The physical space in which people work together will influence their creativity, collaboration, and ability to see and interact with each other. For example, open office spaces were extremely popular for some time, with many of the largest organizations on the planet (including Google) adopting them.

However, while an important step-up from cubicle-only workspaces, open offices aren't ideal in every setting. They've increasingly come under fire for hampering productivity and concentration as well as communication.

Studies show that open offices boost health and reduce stress[14], but at the same time that they may reduce face-to-face interaction and could reduce collaboration[15]. While studies often seem to contradict, a mix of both private workspaces and open areas, with private offices and meeting rooms available is often a happy middle ground.

By creating a mix of workspaces, including private areas, meeting areas, and open-plan desks with flex seating (each person chooses their own desk every day), you give individuals the freedom to work how and when they want. Rather than being forced into a collaborative space or a private space, they can choose. This isn't always possible with the space and building constraints faced by many companies, but it is still important to consider the physical space individuals are working in, how teams are grouped together, and how they can physically collaborate when designing your office space.

Strategy Tip: Create work conditions that promote productivity and performance.

---

14 https://oem.bmj.com/content/early/2018/07/27/oemed-2018-105077
15 https://www.inc.com/jessica-stillman/new-harvard-study-you-open-plan-office-is-making-your-team-less-collaborative.html

Working with remote teams

Remote work is becoming increasingly more common, with freelancing, remote offices, and even entire remote teams becoming popular for reasons of convenience and cost-savings. In some cases, professionals even prefer to work out of the office, benefiting from more freedom, reduced commute, and decreased costs. Many are also more often highly educated, self-motivated, and able to work from home or flex offices without reducing productivity. As a result, it's something that you almost cannot avoid if you want to remain competitive on the job market. Remote work can also be extremely positive for employees, because it helps with creating a positive work/life balance.

At the same time, many organizations are concerned that remote work will affect the quality and productivity of work. Magic happens when people come together and can inspire and feed off each other's ideas. You lose that when everyone is working from home or a remote office and collaborating over Skype or another tool. On the other hand, if you're forcing people to come together when they truly prefer to work from home, they will be unhappy, and you'll lose the "magic" of collaboration anyway. You need to find a good balance where employees are expected and want to do both.

You will have to adapt to remote workers and all the inherent challenges. Depending on how your workforce interacts, this could include time zone, Skype connections, collaboration, and time availability. Preparing for these challenges, whether from allowing in-office employees to work from home a few days a week or working with a remote office in another country, will help you to make the most of every form of work available to you, so that you can get more from your employees, and in a way that contributes to productivity and collaboration.

At Nmbrs, we work with offices in Amsterdam and Lisbon. We regularly host video meetings to aid in communication and collaboration, but this has its own inherent problems. It often takes individuals 5 or more minutes to prepare headsets once we begin, background noise is always an issue, and other small issues which would never be a problem in a face-to-face environment continue to crop up. If someone is busy, unavailable, or even away on holiday, someone in another office might be unaware which can create communication difficulties and even bottlenecks.

We specifically work to overcome issues by creating a balance, bringing people together across offices with digital screens steaming each office space, and creating regular digital meetings to bring people together. Calling people, knowing when they are available or not, and communicating easily are still difficulties which we consistently work to overcome.

Here, creating proper process management and tooling is crucial to ensuring that people can work together well across distances. If people have the same access to tools and the same processes, as well as digital workspaces, working together becomes much simpler. This allows you to more easily work with remote offices, to allow individuals to work from home, or to bring on freelancers and outsource employees in other countries where necessary.

At the same time, working together in a physical space is important for fostering creativity and collaboration when possible. In some cases, working together in a physical space can be essential to the work. Companies ranging from Yahoo to Zappos have taken strong steps to bring employees together, enforcing working together in physical spaces, often in stark contrast to previous lax work-from-home policies. These rules are intended to foster collaboration by forcing people together, where they can talk and interact well. Following policies where many employees worked from home full time, these changes are a strong indicator that not having collaboration simply didn't work for the companies. Unfortunately, new policies of working completely in-office were met with resistance.

Creating a balance where you bring people together when needed but otherwise allow them to work where they want is one solution. Defining which types of work can be handled remotely versus which should be done collaboratively is important. For example, you can decide that work which can be done with minimal interaction can be done at home or from anywhere, but insist that teams come together at least a few times a week for continued collaboration. Similarly, you would want planning, retrospectives, and other types of team-oriented activities to be handled in-person, so that every member of the team can contribute, share, and add value.

While remote work can be extremely valuable, especially for employee satisfaction, you do want to ensure continued team interaction and collaboration. You don't want a situation where no one sees each other or where people are just working but not connecting their work.

When remote work isn't avoidable, tools including video conferencing, cloud sharing, and digital workspaces will help bring individuals together. Even creating intranet or a constant video conference will help. It's also important to consider the cultural and physical location differences between teams. If you're bringing remote offices together, your team's work methods, communication styles, and even languages may not match. It's crucial that you take steps to bridge gaps so that individuals can work together as efficiently and productively as possible.

When people come together, magic happens. I say this because it's always been true in my organization. Nearly every great idea, solutions implementation, or change that moved the company forward, stemmed from an informal environment where people were sitting together. My advice is to make time and space for those moments to happen.

Strategy Tip: Magic happens when people are together, so work to enable such moments, even in remote environments.

Working with team dependencies

As you design teams, you are working to put the conditions in place to increase productivity, to develop new features, or to deliver more and more quickly.

Unfortunately, during real-life implementation, new teams actually slow processes down. Why?

Team dependencies are a near-invisible force that block work by slowing down each team's ability to continue work. Sometimes, when designing teams, you know that one team needs something from another. In other cases, dependencies are less visible, but they still exist. These dependencies force one stakeholder to wait on another in order to continue work, which will slow or even halt productivity.

Direct dependencies are visible and are typically translated to direct software or work-output dependencies. For example, a front-end team needs a back-end team to develop a certain feature in order to deploy a new User Interface. Or, a front-end team needs a DevOps team to deploy a certain infrastructure.

Dependencies of this nature create bottlenecks where one team is waiting on another. However, Product Owners and Scrum Masters can quickly align work and keep teams productive. In the case of very large products, a Scrum of Scrums, where teams get together and align everyone, will be necessary. You also need chapters to align experts, so that everyone is on the same page across squads and teams.

It's also important to pay attention to recurring dependencies. Here, you may want to redesign teams and distribute responsibilities to reduce interdependencies, which will slow work over time, even when managed correctly. Teams should have as much autonomy as possible so they can work without creating overhead to constantly align work with interdependent teams. Restructuring team modules, scopes, or responsibilities into more of a DevOps mindset where teams are developing while handling operations and servers will minimize work dependencies.

Strategy Tip: Adjust teams' scope and responsibilities to avoid recurrent dependencies.

Indirect dependencies are less common than direct dependencies, but also more difficult to spot. Here, teams are waiting on the work or output of another team, but in a way that is not directly linked to that team's output.

How does that work? Let's say you have a team introducing a new technology that will enable other teams to develop their own modules in a better way. Existing teams must decide whether to continue maintaining the old modules or migrate to the new technology. Customers keep asking for fixes and small improvements on the existing modules. But, is it worthwhile for the team to invest in something that is in the process of being redeveloped? When is it worthwhile to stop? Here, the team handling maintenance will linger in a sort of waiting state, able to continue work but uncertain if they should. They are, in essence, waiting on the team to finish introducing the new technology so their work continues to add value.

This problem is exacerbated when teams developing new infrastructure or technology are unaware of stakeholders waiting on their work. These types of indirect dependencies are difficult to detect and will have a big impact on productivity. Communicating with key stakeholders is crucial for avoiding these situations. Here, communicating deadlines, expectations, and progress with stakeholders will keep everyone informed so that teams can make better decisions regarding their own work.

> **Strategy Tip:** Share team strategies and progress with stakeholders to unlock invisible dependencies.

Transfer knowledge to the next generation

As your organization begins to grow, transferring existing knowledge to future team members will become one of your largest concerns. Organizations entering a scale-up stage have to quickly increase work capacity and will onboard new people rapidly. Senior team members are the only knowledge resource, but they don't have the available time to mentor each new addition. Instead, experienced employees are often given the most difficult tasks, which only they have the experience to do well. At the same time, those experienced people are the only people who can truly on-board new hires.

This creates a knowledge gap that is difficult to overcome. The more you grow and the faster you have to onboard new people, the larger this gap will become as your ratio of experienced people to newcomers grows smaller and smaller.

How can you combat this and prevent knowledge gaps as you begin to grow? The easiest and most obvious solution is to integrate a mentorship program from the earliest stages so that a gap never becomes sustainable. Instead, new employees are onboarded with a mentor and that onboarding is a priority for the experienced person.

You can also integrate strong internal documentation so that new team members have access to information and onboarding programs. For example, if processes exist but aren't properly documented, newcomers will have difficulty learning and retaining or maintaining that information.

Mentoring is the easiest way to share organizational culture, values, and real-world application of knowledge inside the organization. Mentoring also ensures that new hires can receive information that hasn't yet been documented because many processes won't be, even when you have a good structure in place.

Culture is often not written down or documented in a tangible way. However, it's an important informal vehicle for transferring information and onboarding new people because it explains how things work, how decisions are made, and even who is responsible for certain decisions. A mentor is the fastest and most efficient way to communicate information like culture.

At the same time, it can be difficult to determine how much time mentors should invest in new hires. Unfortunately, there is no clear-cut answer because it depends on your organization, your current knowledge gap, the person's position, and how long their onboarding time can be. As a rule, the more responsibility the new person has, the longer their onboarding process should be and the more time the mentor should spend with the newcomer.

Creating a mentorship program will also help you to account for other factors contributing to a knowledge gap. For example, if experienced people leave or roles and products increase in complexity. Creating a balanced and sustainable structure for sharing information with new hires will help them to get up to speed quickly so that you can continue to grow without a knowledge gap.

Strategy Tip: Create a mentorship program to onboard new staff.

Succession planning

Small companies often develop around the talent, skills, and motivation of a few key players who do most of the work and bring the organization together. Losing any of them at an early stage would mean losing the organization. However, as you grow, continuing these same dependencies is an often-fatal mistake.

People will always leave and being overly dependent on any one person, to the point where you can't replace them, will mean huge setbacks when they do leave. This is difficult to avoid in the case of the CEO or another founder, but very fixable in the case of key designers or technical roles. If your head developer quits tomorrow, where will that leave their team?

While dependencies will naturally fade as your organization grows (most companies can only truly afford to lose people after reaching 80-100 people), it's also important to think about succession planning and replacement before that point.

Succession planning is the process of either developing internal people to potentially fill important roles or to have policies in place which prevent important people from leaving without finding a replacement. In both cases, this process is difficult to implement before you reach that 80-100 people marker, simply because many of the people on your team will be filling several roles.

Succession planning involves a simple process of a) defining which roles are important and their impact when not filled, b) defining how to fill those roles should they empty and c) determining who is responsible for filling those roles.

In most cases, succession planning should primarily be an HR concern. HR will best understand who is who in the company, which existing employees show potential for development and promotion, and will better be able to manage succession planning during performance review. Aligning with HR to ensure that a succession plan is in place is a crucial step. People will always leave and you need to have people ready to take their place and responsibilities. Working with HR to develop a pipeline is very challenging for a small organization but is necessary.

HR can perform a data analysis to check churn and predict how often you have to hire. You can use industry and benchmark standards to predict how long individuals are likely to stay in your organization. And, you can work with HR to line up competent people to potentially fill positions and to begin hiring in advance. Doing so also allows you to allocate budget that meets needs based on what is likely to happen.

You can implement numerous strategies for succession management. Facebook requires that all employees know who their successors are, so that if they were to leave, the team doesn't fall apart. This policy has helped fuel Facebook's rapid growth, because the company has never had to worry about employee movement. Similarly, you should know who your key players are and who can or will replace them whenever possible. Of course, very small organizations with only a few employees will struggle to implement defined succession planning, but it is something to consider implementing as early as possible.

For example, Ram Charan, author of the Leadership Pipeline, proposes that good leaders and technical leaders should be grown or fostered inside the company as often as possible, because internal people understand culture and work processes. This worked very well with Microsoft's Satya Nadella who used his existing understanding of Microsoft's core culture to change it for the better.

Integrating a similar internal policy (whether you make individuals responsible or not) will allow you to understand who can move into existing roles, especially key roles, should they empty, and why.[16]

Recruiting is time-consuming and will cause significant delays if you must replace someone in a key position. Recruiting internally, creating a succession plan so that someone is always ready to fill an important role, and working to develop existing employees based on skill and potential will help you to stay productive even when key people move on.

> **Strategy Tip:** Understand your key players and who can replace them should they quit.

While it's important to understand succession planning, you also have to plan for when people are simply not there. People may be absent from teams for several reasons ranging from holidays to sickness or other emergencies, and you must be prepared to let them go without hurting your team's ability to function and produce value.

Here, it is valuable to introduce a buddy system, where each person on a team has a "Buddy" who can take over and keep things running while they are gone. Everyone should have an assigned and visible buddy who can take over key responsibilities during absences to reduce dependencies.

---

16 https://www.businessinsider.com/mark-zuckerberg-succession-plans-performance-reviews-2017-8?international=true&r=US&IR=T

For example, if someone were to call in sick during a crucial moment in a project, it's important that they be able to do so. It's also important that the team have someone who can step in and take on key responsibilities rather than causing a delay.

Strategy Tip: Make sure everyone has an assigned buddy and everyone knows who that person is.

Monitoring team health

Designing teams in an Agile environment means building them to create as much self-sufficiency and autonomy as possible. At the same time, you want to be able to manage how teams are performing so that you can keep track of output, team capabilities, and productivity.

You can achieve this by defining relevant KPIs to track, measure, and define what value is for the team so you can measure where and how teams are contributing. You also want to measure velocity so that you can see how much work teams are putting out. However, team KPIs will vary a great deal depending on the team and their purpose.

One easy mistake here is creating a monitoring process in which teams are not aware they are being monitored. Failing to communicate your performance monitoring process to teams will impede their ability to improve because they aren't aware of how they are being measured. It also fails to build trust and may even deteriorate trust if teams discover they are being monitored.

Creating a simple health-check is one easy way to establish team involvement in performance monitoring. Defining HR, operational, and technical topics aligned with team goals and output gives you a platform for self-assessment, because teams can state how they think they are doing on each topic. You can combine these topics with data involving velocity, support tickets, and other KPIs to develop a complete picture.

For example, at Nmbrs, we use a system of Green, Yellow, Red health checks:
- **Green** - Everything is okay, keep monitoring and take no action
- **Yellow** - There are problems, but the team is aware of them and are working to create a solution. Here, it's good for the team to solve their own problems
- **Red** - There are problems and the COO must step in to take action. This problem may be structural and may require changing the scope or purpose of a team or splitting the team

Monitoring guidelines must encompass the scope and purpose of the team, how it contributes value, and velocity. This will allow you to see if the team is really performing, where it's going right, and where it's going wrong. This type of monitoring is most often performed in collaboration with a Scrum Master or a person in a similar role, because this person is responsible for keeping the team moving forward.

Performance monitoring should be handled frequently, and you should integrate both small checks after projects and larger ones at standard points in the year, such as on a quarterly basis.

> **Strategy Tip:** Define what makes teams successful and monitor them to ensure continued output.

Team satisfaction vs customer satisfaction

While the logical goal for any organization is to strive for customer satisfaction – because that's what brings value – internal employee and team satisfaction are equally important. Both are among your most important metrics, but both play into each other in ways that mean you can't have happy customers if you don't have happy employees.

How so? Customers actively notice when team satisfaction drops. Unhappy teams do not produce as well or as much, are not motivated to create innovation or new products or features, and often don't really care how the organization fares and therefore don't care how the customers fare.

Happy employees logically translate into happy customers. If you only focus on customer happiness, you won't necessarily achieve it, because your employees may not be happy.

At the same time, employees often want different things than customers. If customer demand for a new product or feature is very high, you may put a lot of effort into pushing teams to deliver quickly. They may have to work overtime or put other projects on hold to meet a commitment to the customer, because you prioritize that value. Doing so will almost always reduce team satisfaction, because they might feel too pressured and stressed, may not agree with the goals, or may simply prefer to work more interesting code. Everyone likes to develop new "cool" stuff, but we often forget that, in a SaaS company, it's not all about coding, testing, and implementing new things – it's often about simple maintenance and keeping existing software running. Unfortunately, this work can be boring and demotivating and you can't ask teams to dedicate 100% of their time to it.

The more you push, the more it will backfire, because your teams will become more and more dissatisfied and unable or unwilling to deliver real value to the customer.

Creating a balance between customer wants and employee wants is important if you are to maintain the status quo between the two. Here, you have to consider employee and customer goals, align them as much as possible, and work to highlight mutual goals and reach compromises where necessary.

For example, Google encourages employees to spend 20% of their time on new ideas. This initiative gives employees the freedom to work on things they want, alongside "boring" work. In Google's case, it's also resulted in new products like Gmail and Google Drive, creating value for Google and for the customer. Balancing demand for product maintenance with employee's need for creative and interesting work will drive more business opportunities in the long-term.

Agile practices often make balancing customer and team satisfaction significantly easier than waterfall. For example, if you can identify and communicate overarching goals for the customer to teams, they can often work much more closely in line with what the customer wants, but at their own pace. You can also take steps to identify obstacles and problems, such as too-tight deadlines or a lack of tooling to help teams align themselves with those goals, without adding pressure. Giving teams space to work on things they want to do is another way you can improve motivation and satisfaction. Why? Even though work is not a "reward" in and of itself, most people are happier when they can choose what to work on, even if it's not the most important thing in that moment.

It's important to create mechanisms that measure team happiness. Unfortunately, "Happiness" is very much a variable thing and it's difficult to create metrics for it. Instead, I recommend connecting with squad leaders on the floor and asking. You can also implement tools such as surveys, regularly actually walking through squads to ask how people feel about their work, and otherwise specifically asking rather than attempting to gauge happiness through vague metrics.

There are also many existing frameworks and services that exist for the purpose of measuring and gauging team happiness or satisfaction. These vary in cost and value, but largely consist of preexisting sets of surveys, metrics, and KPIs, which are then tailored to your team.

Why? It is very difficult to gauge happiness based on metrics, because teams often perform even when poorly motivated or satisfied. At the same time, it will pay off to understand how your teams are doing because you can manage them to reduce churn and ensure long-term productivity and motivation. You can integrate this as part of your retrospective cycle but should also measure how teams are feeling throughout sprints and work cycles.

How did I make this work in my own organization? We created a self-assessment health monitor, where teams can rate themselves on topics such as vision clarity, technical quality, motivation, workload, and team dynamics. Our Agile coaches then collect this data and work together with the team to improve their performance and happiness.

Strategy Tip: Balance customer and team happiness.

Conclusion

Team design, structure, and organization will greatly impact your organization's ability to work, grow, and drive value. As a result, team performance and management are essential elements of operations, but elements which you will share with HR. These processes are fluid and ever-changing, because team design must meet the needs of the company, its strategy and vision, and its people.

Building a close cooperation, keeping HR in the loop, and collaborating regarding operational needs will help you to do more with your teams. People form the backbone of your organization. They produce the labor, ideas, and ingenuity which result in both direct and indirect value for your customers. Managing people, in ways that benefit those people, will help you to build a better and stronger organization.

Data and Information Management

"Data is like garbage. You'd better know what you are going to do with it before you collect it."

– Mark Twain

Data and Information is the final pillar of the Vision to Value Framework. Data is a unique part of the model because it functions less as a standalone aspect of operations and more as the "glue" that brings everything together to create a cohesive whole. Collaboration and coordination are essential in connecting your organization and data is their driver.

Why? If you have data but don't use it, the related aspects of operations become useless. If you have a company vision but don't share it with the organization, it loses value. If you invest time in describing processes but employees are unaware of them or cannot access or use them, those processes achieve nothing. If you organize work but it's not visible and teams don't know what others are doing, you can't easily coordinate large projects. If teams or individuals are unaware of who works in which role and what their responsibilities are, maintaining compliance and auditing for quality will be difficult-to-impossible.

The smaller your company, the less you need data to bring everything together. As you grow, it becomes more and more crucial to every aspect of operations.

Communicating data and information is what brings organizations together across projects of any scale, including day-to-day operations and maintenance. This communication can take many forms, ranging from collaboration tools such as Slack, Hangout channels for calling, intranet, project management tools, and documentation tools, KPIs, and visible work or methods. The important aspect is that people are connected so they know where the company is going and how, how teams are performing, and what each team is working on.

Defining a proper data sharing strategy ensures operations has the communication tools to continue running well, because each member is aware of what is happening, who is doing it, and why.

Work Connected

While organizations are built and structured in many ways and deliver many different types of products, all of them bring people together to work. Making that work transparent across the organization, rather than simply to interested stakeholders, is crucial to enabling collaboration and information sharing across teams.

If you have two teams working together in the same room but who don't collaborate, you could have an instance where Team A documents a process and Team B is working on a similar project and creates their own process which almost exactly mirrors the one developed by Team A. Creating a communication standard between the two would have saved both work, because they could have collaborated.

Creating these channels means understanding how people work and designing tooling supporting individual teams in how they collaborate. Some teams share video, diagrams, source code, or might have to talk or share their screens. Each will need their own solutions. Even if people are working remotely, they need to be able to sit down and communicate in a way that makes them feel connected.

Understanding how people work and developing channels that meet their needs will enable them to collaborate as efficiently as possible.

Creating quality communication channels can help you connect people in ways that even working together in an office cannot. Many people believe working remotely is a bottleneck, but it rarely is. Instead, improper communication or improper collaboration channels, which prevents or fails to enhance that collaboration often becomes the bottleneck, even inside of offices.

Working in the same tooling is one of the most efficient ways to force communication and collaboration, because everyone works in the same way, using the same features, and the same processes.

Work, especially building software, cannot be completed alone, and it's up to you to create channels for people to communicate and collaborate.

With more and more employees working remotely, it's also important to consider remote collaboration. Here, lack of physical communication and collaboration forces you to create very clear processes for communication to ensure that effective channels exist and are followed.

If you need to get in touch with any member of the team or vice versa, you should have a defined way to do so such as via text, email, or a phone call. You should also have structured communication method for team meetings, such as in-person or over a video conference call with regular communication inside Slack. Defining how communication is handled prevents issues such as receiving important information on a range of channels, with no set place to check for team communication.

Finally, it's important to define when you will communicate. This is most obvious when setting up team meetings, where you will likely have an obvious and set time, such as "every Monday morning at 9 AM". However, you also want to define regular team communication. Are team members allowed to communicate after hours? If team members are in different time zones, how can you prevent issues with personal space and privacy? Establishing communication norms and standards for how long individuals can take to reply is also important.

Each of these aspects of communication becomes even more crucial when dealing with remote workers, because removing physical collaboration can make other forms of communication even more difficult.

Strategy Tip: Ensure people can easily reach each other to collaborate.

You'll notice that many people collaborate in different ways and using different methods. This can result in major communication problems. Here, it's important to observe how people interact and connect. If you find impediments, you should address those with a solution.

In a hypothetical situation, you design your organization, put processes into place, allocate teams, and you logically expect everything to run well. Then, you notice this isn't the case. Bottlenecks are everywhere, with teams waiting on each other, on management approval, and on task lists. When you ask what's happening, people say things like "I didn't know".

Scenarios like this one don't necessarily mean processes aren't designed properly or that people don't understand their roles or responsibilities. They're often simple communication problems.

While there are many reasons for communication problems inside of organizations and between teams, avoiding them should be a priority. Communication makes things happen and it's essential to your organization. If the right stakeholders aren't informed, a project might as well not exist. Communication is also different than information, because if people aren't aware of information, it might as well not exist. Developing the proper channels to ensure information is broadcast and that it reaches people is crucial to creating that communication. This is important to operations because failures and impediments in software development are often related to communication failures.

In a real example of communication issues I've experienced myself, my company outsourced a product to an external developer. At one point, we received a bad implementation. I called the organization to discuss the problem with the front-end developers and discovered that they weren't aware of what the back-end team was doing. Instead, they made an assumption which resulted in a bad implementation.

Internal communication should always be your focus when developing channels, because external communication is less important. Why? External communication surrounds marketing and external investors while internal communication ensures everyone is aware of what they are doing, what each team is responsible for, who is in each team, escalation processes, etc. It's essential for collaboration of any kind.

Good internal communication is about messaging, talking, and sharing assets and documentation. It's about communicating at an organizational level, sharing company news, product updates, company vision, and important events so that everyone knows where the organization stands.

There are many reasons that communication issues arise. For example, team members may be convinced of their knowledge of the other, there may be no transparency in work, some individuals may be failing to listen to others, or you may have issues with hierarchy or cliques. Each of these can be combated by promoting work transparency, promoting a culturally diverse work environment, and pushing multiple forms of communication across your organization.

You also can't blame all internal problems on communication. While it's often easy to turn to communication problems as a likely culprit for issues, communication isn't always the problem. Instead, poor communication can be a symptom of a deeper problem such as a negative work environment, poor company culture, no clear definition of roles or responsibilities, lack of organizational structure, and so on.

Good communication will feel like oil in work processes, it will make everything run more smoothly. However, if processes aren't good, communication issues will arise as a result. If you're having communication problems, it's important to review the problem, determine when and why they are cropping up, and take action to either improve communication standards or to solve the underlying process issues beneath them.

As a Tech COO, it's important that you put communication channels into place, ensure they are accessible, and hold people accountable for using them.

Strategy Tip: Create structured communication channels and tie them into work.

Communication is a two-way street

While organizational efforts towards facilitating communication are geared towards individuals and teams or broadcasting information across the organization, it's also important that individuals be able to communicate back.

For example, it's a very common process to broadcast emails. The thing is, broadcasts aren't communication. In fact, most organizations see an average open rate of about 69%[17] for email broadcasts.

---

[17] https://gethppy.com/employee-engagement/infographic-the-benchmark-of-successful-internal-email-campaigns

If individuals can't communicate how they are receiving information, you can't ensure they understand what is being shared or that they are receiving it at all. If you simply broadcast information, people aren't saying they understood the message or the communication method. Taking the time to see which types of communication work for your teams is one important element of internal communication.

If you can request feedback on information sharing and determine that emails aren't being read or understood because most people on the team receive too many emails, you could invest in a Slack channel or video for important updates to increase engagement and understanding.

Communication is a two-way street and it's important to listen, look for information, and to truly understand what is being communicated. If someone on a team is saying "I wasn't informed of that" when it was previously shared, you need to review how you're sharing important information and redefine your communication channels.

While some organizations attempt to combat this by marking everything as "urgent", this will eventually backfire. People will stop paying attention and won't listen when something important is shared, much like in the story of Peter and the Wolf.

Leaders assume people know things, but they often don't. Information is lost when broadcasted, when handed off between teams, and even during onboarding.

Creating a clear communication structure, which defines how, when, and where information is shared, is important. Individuals should know where to look to find information if they missed it the first time. You can work to solve instances of "lost communication" by creating clear processes defining "how to communicate" in important instances such as when onboarding, when starting new projects, when teams are sharing or communicating work, for processes, and so on.

You also want to define information maintenance processes, so that information stays up to date, individuals always have access, and everyone knows where to find it.

> **Strategy Tip:** Don't assume that shared information is acknowledged, find ways to know how messages are received.

Internal communication channels

Internal communication should help you share information across the organization, but it should also facilitate team communication. Team members should be able to share information within the team through their own platforms and channels without broadcasting it across the organization. This is very useful for keeping conversations within the scope of a team without sharing it out of context.

Here, it's extremely useful to integrate tools such as team-wide chat or platforms like Slack where teams can message, call, and share files within a team group. However, you also need to consider the team's technical requirements and preferred communication methods. A marketing team won't likely need much beyond chat or video calling. On the other hand, technical teams will likely require screen and diagram sharing as well as easy ways to collaborate on technical diagrams, functional designs, or source code.

Developing individual channels to meet the needs of each team is always a good idea, because each team will have its own technical needs. How do they collaborate? How do they handle product updates or critical escalations? Can channels handle triggers to notify responsible parties when either happens?

While you (obviously) don't want to invest in an endless range of communication platforms, it's important to add enough that most teams can communicate in a way that meets their needs.

Most teams will need email, chat, and video at a minimum, and will use all three. Additional communication methods may be necessary or essential depending on the team, how they work, and their technical needs.

For example, chat is a new standard for inter-team communication. Many people even prefer it to calling because it requires less effort in that you can chat instantly without headsets, dialing a number, figuring out Skype or hangouts, etc. This is especially important when people don't have access to easy and well-understood ways to call, because chat is an almost effortless alternative. Chat is also less disruptive than calling in that individuals don't have to stop what they are doing to communicate. However, chat tools can be extremely disruptive because they can result in too much messaging. How? If everyone is on the same platform and constantly sees messages, it's difficult to keep up and information overload becomes a thing.

We integrated a chat platform at Nmbrs to facilitate better communication. Unfortunately, it doesn't always help because individuals receive too much information and it often isn't relevant. Instead, they're distracted by things happening in other teams or even just casual and personal chat which can be disruptive to work.

The only real way to combat this is by creating private channels and chat rooms for individuals and teams so they don't see everyone's messaging. It's also important to ensure that everyone knows where their channel is, which topics are okay in the channel or chatroom, and how and where to go to communicate with other teams. If you have a general chat, it will be overloaded, and you cannot easily communicate large amounts of information to a team.

At Nmbrs, we defined topics including an email list, a wikispace, internal teach communication for each team, inter-team communication, individual processes, a code red channel, and so on. Defining topics for individual channels and rooms allows us to keep communication on-topic so that when something happens, everyone knows and it's relevant. Inter-team communication stays in the team, team collaborations are shared in a relevant space for when teams need to know when updates are happening or if they are working, and if there is an escalation, the code red chat is never flooded with irrelevant information, so everyone immediately knows something is wrong.

It's also important to keep in mind that not every communication channel is ideal for everything. We used to have office-wide video conferences between our offices in the Netherlands and Portugal. However, we quickly discovered that teams would often have inter-team conversations, leaving the other team out. Here, one-on-one video calls were much more efficient, because people could better understand and communicate with each other.

This doesn't mean that office-wide conferences can't be valuable – they can – especially when trying to get everyone on the same page and when introducing new people. However, they may not be as valuable for important work items like cross-team collaboration, where everyone needs to feel included.

Channel	Suitable for:	Not Suitable for:
Email	External communication	Internal conversations with interactions. Here, it's better to use chat
Phone	Quick internal discussions. It's still much faster than video calling.	More complex collaboration. Consider video calling instead
Chat	Simple conversations. Involve several people around a topic.	Complex discussions, problem solving, crisis situations
Video Calling	General-purpose, remote collaboration, interviews, meetings, etc.	Replacing physical interaction (requires too much equipment such as headsets, etc.
Physical Interaction	Most collaboration. This is where magic happens	Remote collaboration
Project Management Platforms	Sharing project status with teams and stakeholders	Maintaining conversations across teams
Intranet	Broadcast information, memos, documentation	Important messages

Choosing the right platforms for communication will help you to communicate more smoothly and in ways that are more appropriate for the work being completed. In most cases, teams should be involved, because most will have their own preferred ways to communicate.

Strategy Tip: Establish clear communication channels matching the work at hand.

External Communication

While internal communication enables teams to work together and produce products, external communication enables teams to work on products and features customers actually want. Here, you must establish external communication with external stakeholders, customers, and end-users.

Why? When teams are isolated from customers and end-users, they often build products and features they believe are important. These features aren't necessarily the same items the customer would prioritize. Without communication, you can eventually realize a large gap between customer needs and produced features. Ensuring that product development teams have an open channel to receive customer and end-user feedback and demands is the best way to create feature-validation, especially when team members are not product users themselves.

Strategy Tip: Establish frequent communication with external stakeholders.

Here, one easy solution is to create external channels that communicate a roadmap of features and intended development to customers and end-users. Their feedback and input will provide development validation.

Most SaaS products have a large number of customers and it's not always easy to connect with individuals directly. A good approach is to develop user groups and personas, such as "expert user", "starter", "IT manager", "compliance officer", etc.

You can then develop a forum or another communication platform where Product Owners can either see and review customer feedback and interaction or communicate with them directly.

What should you be communicating? I believe it's important to have channels for product updates, known bugs and incidents, questions, and general feedback. The more you communicate to customers and end-users, the better. Transparency in product development will increase trust levels. Most importantly, communicating bad news is always better than no communication so you should never be afraid to share "bad news" updates.

Considering which channels you will use to communicate information externally to stakeholders and users is equally as important to creating dedicated internal channels:

Channel	Suitable for:	Not Suitable for:
Email	External communication	Scaling to standard information pushed to all users, such as features from an update. Use a forum instead.
Phone	Quick and efficient talks, Customer Support	More complex collaboration. Consider video calling instead
Chat	Customer support, Sales assistance	Complex discussions, problem solving, crisis situations
Video Calling	General purpose remote collaboration, Onboarding	Replacing physical interaction
Physical Interaction	All kinds of collaboration.	Remote collaboration
Knowledge base	Update notes, Self-help or FAQ information,	Guided self-help, Customer service, Sharing frequently changing information
Forum	Product topics, releases, software changes, customer interaction	Important messages

Strategy Tip: Setup platforms to communicate with end-users and customers directly.

Internal Documentation

Internal documentation ties into nearly every aspect of operations, enabling processes, communication, quality assurance, onboarding, and much more. Documentation is the process of recording processes, knowledge, projects, work completed, and decisions made, so that it is accessible and available in the future and for future employees. Technical documentation helps teams decide what to build, why they have to build it, and how to do so. This enables individuals to complete processes or onboard new employees that much more quickly.

If you begin a project with proper documentation, you can see which decisions were made, where, and why. You can later use this information to re-do the project or to recreate it, to train new employees, for quality assurance through process documentation, and to communicate the progress of the project to other team members or stakeholders.

Good documentation also means you can answer questions more readily. If someone asks a question and you can link to documentation instead of creating a new answer, you can add value and share detailed information without the fallibility of human memory.

Documentation is valuable and should be accessible across your organization. For this reason, it's always a good idea to invest in a proper internal documentation platform that everyone can access. You can create this in the form of a wiki or intranet or even a custom platform designed for the purpose. Intranet was previously the most popular solution for internal documentation but it's falling in popularity because of technology gaps. Today, one of the best solutions is to create or develop an internal platform or central documentation platform where teams can add their own content including processes, vision, squad information, and general processes and documentation.

No matter what you use, your platform should be searchable, should offer a common structure to everyone, and should be accessible to everyone. Teams should also have their own spaces inside the platform, but those spaces should use the same structure as the rest of the platform.

A documentation platform should include team spaces and project documentation, but should also have a searchable knowledge base or documentary for general organizational information such as work dynamics, office and locations, HR policies, processes, calendar, teams, responsibilities, roles, and so on.

This type of documentation knowledge base will aid in onboarding, process review, quality assurance, and when moving projects between teams. Individuals can more easily share information, decisions, and progress because the documentation already exists.

Ensuring that information is always written down will greatly improve onboarding processes because individuals will have written and documented processes and projects to review. Not only can they see what needs to be done and what is expected of them, they can review how previous projects were completed, what decisions were made, and what the results were. While much of this information will be shared orally by a mentor, having an official and documented version is always extremely helpful for clarification and memory. As your organization begins to scale, documented information also reduces the load on onboarding mentors, because more information is available to simply be shared rather than taught through real-world experience.

Nmbrs integrated a wiki platform which we call the Nmbrs Book. The Nmbrs Book is a centralized information platform, containing everything employees need to work in our organization. This includes processes, roles and responsibilities, teams and team members, HR policies, and more. The goal was to create a central repository of information, where any repeatable employee questions can be answered. For example, when someone asks, "How do I apply for a vacation?", "What do I do in case of a code red?" or "Who is responsible for maintenance on the mobile app", the answer can always be, "Check the book".

Having centralized information enables people to access anything they need without waiting for an answer, relying on human fallibility, or creating bottlenecks. Because it's a Wiki, individuals can update information when something changes, enabling us to assign content ownership to relevant teams and Product Owners. As a result, it's driven a lot of value for us in terms of improved communication for documentation, processes, and ownership.

The larger your organization becomes, the more critical having proper documentation becomes. As an early stage startup, you can likely get away with little to no documentation. However, documenting early decisions and processes will give you valuable insight into your early decisions when scaling your organization. It's also crucial that documentation be organized in such a way that it's clear who owns it. One of the biggest drawbacks of documentation is maintenance, because outdated documentation is worse than useless, it can be harmful when people follow old processes. Ensuring that you always know who oversees documentation, who should be updating it, and whether it's being used is an important part of keeping it up to date and alive.

This ties into process documentation from earlier in this book, where I discussed the need to minimize process documentation by only documenting processes which are necessary and valuable. If the process is relevant and useful and having it on hand improves efficiency or reduces time expenditure, it's valuable to have. If not, you probably don't need it.

Documentation should never be a dead end. It should always be part of a process or flow, where it is updated as processes change, processes change because of documentation, and individuals consistently search and use the documentation. If documentation is no longer relevant, no longer searched, or becomes outdated, it likely means that the necessary trigger of it adding value is no longer there.

Even if you are running a one-person organization, documentation should be built in because it will be useful. You can always look back to see what you thought or did, which will be practical to have from day one. It's very common for teams to work on small projects and skip documentation because you don't have to communicate information to multiple people. Unfortunately, this can backfire. For example, when someone leaves and doesn't document a project, their work and knowledge is gone.

It's often the case that an employee's last task is "please document everything you did". Unfortunately, documenting after the fact often results in errors and liabilities. If you establish a strong process of documentation from day one, this will never be the case. You'll also always have a way to look back to see what everyone did and where it might have gone wrong (or right) so that processes can be improved, duplicated, and passed on to the next generation of employees.

Strategy Tip: Integrate documentation into process flows, so that it stays alive and up to date.

Drive with data

Data is increasingly available as customers move online, processes are completed by computers, and new and better collection methods enable even tiny organizations to collect massive amounts of information on customers, markets, and product development. At the same time, many organizations are failing to implement and use available data to drive their organization or its movement.

While 79% of companies involved in a 2018 survey claimed to see the value in using big data for strategy and organizational development, only 1/3rd of organizations do so[18]. Why? While the answer is complex and often related to a

---

18 http://newvantage.com/wp-content/uploads/2018/01/Big-Data-Executive-Survey-2018-Findings-1.pdf

combination of factors such as not understanding what to do with information. It also relates to moving data from collection to implementation.

This gap between seeing value and utilizing data to realize more value is extremely important because modern analytics can help drive every area of business. Data and analytics enable key performance indicators for tracking growth and performance. They also enable you to decide what markets want and why so you can develop products, gauge customer satisfaction, and even track performance against competitors.

Data should be the basis for decision-making including formulating strategy, vision, team planning, and every other part of the organization. If you don't know which direction your organization is growing, you can't begin to expand the right teams, can't invest in the right modules and products, and can't set an appropriate strategy for your organization.

As a startup, data is the key to fueling growth, because it allows you to make strategic decisions to grow in the right directions. Startups often shoot from the hip and manage because there is so little to manage. As you grow, data structures become crucial to growth. You need them to guide individuals, teams, and the organization.

Here, data doesn't have to be about large-scale projects. It does and should be about developing and defining KPIs at every level (organizational to team-level) so that you can track the performance and progress of goals at those levels. High level KPIs show how you are progressing through your business plan while lower-level KPIs show how teams are performing.

In addition, data availability isn't just about creating high-level strategy, data empowers teams. In a hierarchical team, the manager pushes decisions and only she needs data. In Agile teams, everyone needs to see data so they can make their own decisions based on relevant factors. Here, Agile is a lot about learning and validating, so sharing data helps validate processes to ensure you're going in the right direction.

Driving with data means using data and vision to move the company forward. You use analytics and KPIs to formulate decisions, measure your success, and to redefine your strategy as you grow. It is quite simply the process of using measurable facts as a basis for your decisions. For example, if you use data to define baselines, you can improve based on set and specific metrics. If you don't, improvement will always be based on a feeling.

Analytics and KPIs define where your organization stands and using them is essential for continued growth. Knowing where you stand and how to measure growth gives you the tools to set and define strategy, to measure your success and actions, and to measure every aspect of your organization so that you can control growth and direction.

Strategy Tip: Ensure data is available to support decision-making at every level.

Data-Driven vs. Data Informed

While data is extremely useful, it's crucial to avoid becoming too caught up on using analytics and numbers. Data-driven decisions take people out of the equation, they are just numbers, and you might miss key aspects of organization growth.

If you base all decisions on data, you might miss new trends, changes in customer opinion, or even what developers are doing on the floor. Staying involved and meeting customers, talking with developers, and understanding how the organization is running is important if you want to continue to use data well. You can, however, use pure data for budgeting and sharing information to stakeholders. The bigger your decision, the more important it is that data should play a large role, but it should almost never be the sole deciding factor.

Why? You will never have all the data you need to make a decision. Your data is based on your customers, your demographic, and everything you have access to. You can't completely predict market changes, demand outside of customers you have access to, or shifts in trends using data alone. A data-informed approach means recognizing this limitation and utilizing other resources to make decisions such as targeting new demographics, moving into new markets, or adding new features not currently demanded by consumers.

Data will always be useful, but it does have its limits. You will have to innovate, take bigger steps, and make assumptions or take risks. If you play it safe and only move where data allows, you will limit growth. Instead, data should be used to measure and validate assumptions, rather than as a sole driver.

With this in mind, it's important to pay attention to what data you are collecting and why. Are you using data to measure progress? Or to validate an existing assumption? Or are you using it to formulate your strategies and plans. It's easy to collect data relating to onboarding, site traffic, or number of users. It's much more difficult to use that data to predict long-term customer satisfaction, impact of features, or long-term tradeoffs. Unfortunately, you can't get those answers with data alone. You need a base understanding of your customer-base, what users are doing, and the market.

While any type of information, including what you learn in your organization rather than through numbers, is data, it's important to base decisions on a rich range of sources, such as interaction, market knowledge, KPIs, performance, and even creative and innovative ideas rather than simple metrics. Taking this "data-informed" approach helps you to avoid the pitfalls of using formulas to make a decision.

For example, some organizations use decision-making formulas to determine if something should be developed. Here, a dev team might have to validate that an update will increase conversion. Then, they might have to use A/B testing to ensure everything works towards that goal.

The problem with this data driven approach is that it makes innovation difficult or even impossible. Data will never back up the massive leaps required for innovation. Data-driven is very applicable in controlled environments like maintenance, where teams don't need freedom, but is extremely limiting in other circumstances.

My advice is to create both data-informed and data-driven teams. It's more costly to train people to make decisions using data than to force them with a data-driven formula, which is why you shouldn't invest in it for every team. If you focus on integrating a data-informed approach for innovation or "speed-boat" teams and a data-driven approach for oil-tanker or "maintenance" teams, you should have a good balance of both.

> **Strategy Tip:** Balance data-driven teams with data-informed teams to enable continued innovation.

What to measure?

It's easy to collect data for the sake of data. Some metrics can even seem important or look impressive. If you're collecting metrics such as total registered users, you can likely create a good impression of how well your software is doing. But, do these metrics reflect anything valuable that you can use? It is a good idea to understand how many active users you have, but in most cases, information such as how product improvements are working, customer satisfaction, needed features, and so on, are much more important.

The most important metrics you can collect are those that will help you validate improvements, products, and maintenance so that you can push growth, ensure customer satisfaction, and develop new features.

Does this mean you can't collect vanity metrics? No, it doesn't. They can be useful when sharing success with stakeholders and external people. However, you shouldn't act on them. Your most important metrics will align with business or product goals and will enable you to take direct action.

You'll also quickly find that if you simply collect everything, you'll have more data than you know what to do with. If you collect actionable data, you'll have a leaner view of what is working and what isn't, because you're only collecting the information that matters. This will sometimes require careful consideration into what is or is not actionable.

For example, you could measure the number of bugs in your product. However, this metric isn't relevant because it might not mean anything. Your product will naturally generate more bugs as product size and features increase. When teams perform a bug-bash session and find issues, the bug counter will increase. This metric only becomes relevant when you break it down into internal vs. external bugs. This metric will indicate how users perceive the quality of your product, because it shows what percentage of bugs they see. On the other hand, if most bugs are found by teams and you fix them before customers see them, customers see a high-quality product.

Quality and velocity are two of the most important KPIs you can measure in product development. While there are other metrics, quality and velocity will tell you how much value you are delivering and how quickly. Defining these metrics will give you the tools to improve the speed of your value stream while helping you to deliver more quality with each update because you will understand the factors contributing to each. One common practice for creating quality KPIs is to measure story points delivered per sprint or time frame/period. A story point is a unit dependent on the environment or the team. It will enable you to measure team or value stream velocity, because a story is connected to goals and direct value. Like measurable output, that velocity is not necessarily linked to end-user value, but it will give you a better idea of the team's capacity to deliver.

Quality metrics should include factors such as feature acceptance and customer satisfaction. You can also measure support tickets per customer, what support tickets are related to (e.g. actual problems instead of how to use), and adoption rates.

Velocity rates are often simpler and typically relate to time to update from the smallest change (single line of code).

Carefully defining what you want to measure will give you guidelines to collect data that actually matters so you can act on it.

Here, some better alternatives to replace commonly collected metrics:

Metric	Why it's not a good metric	Alternatives
Number of support tickets	It's difficult to act on this information. An increase could reflect different data such as in increase in customers or an increase in problems with the software	Tickets per customer - This clearly communicates how often customers have problems, directly reflecting on quality and ease of use.
Users	This metric can be misleading because it doesn't differentiate between active and inactive users.	Active users - Specifically tracks to the number of people using your product
Number of story points delivered	More epics may not directly relate to value for the customer. You can track how quickly you're delivering, but what are you delivering?	Maintenance Story Points - Tracks effort invested into maintaining the product Technical Story Points - Tracks investment into reducing technical debt or new features Together, these two metrics show how quickly you are delivering new features (perceived as value by users) or internal improvements.

Strategy Tip: Define actionable data and avoid collecting data that doesn't matter.

Once you've defined actionable data, you have to work to act with it. This means defining which data is important, which targets and triggers create actions, and how individuals and teams must react when those metrics are triggered.

Here, you begin by defining targets and triggers to determine actions. This means defining key metrics that correlate to important business goals or strategy and creating triggers for them.

Once defined, you can set a target of either a minimum or maximum KPI and define an action for that trigger. For example, you can create a trigger if the number of bugs per week goes above a certain ratio or amount. Once that trigger is reached and the number of bugs is too high, teams could be triggered to organize special sessions to fix the problem or review QA to determine where problems are occurring.

The first step is to define baselines for your metrics. What is normal behavior? What is current behavior? What is desired? If you know where you are, where you want to be, and what's normal or standard, you can begin tracking that to what the team is doing, how they feel, workload, quality of performance, etc.

You can then use your baselines of normal to setup triggers. If team performance is low, individuals are working too hard, or quality drops, you want to know. If you know what counts as "too low" or "too high" you can create a trigger to set up an action on a team or organization level to remedy the problem once it occurs.

What should those actions be? It's important to agree on targets with teams. These targets should be realistic and achievable. For example, if teams are seeing too many bugs per week, you could set a target of decreasing bugs by 20% for the next month. This is much more realistic than something like "zero bugs starting next month" and will likely achieve results, where the unrealistic alternative most likely would not.

> **Strategy Tip:** Design processes to create actions when specific KPIs are triggered.

Collecting data

Data is a crucial step in the process of operations, but where does that data come from? The answer is often, from your existing tools and processes. Most modern organizations create and collect massive amounts of data through existing tools and processes. Some of these tools even collect and share data in dashboards and reports, enabling individuals to review everything from efficiency to success metrics as defined in the tool.

However, it is important to bring all your data together to create useful metrics. While you don't need all data from across the organization in one place, teams need to see their own data, decision-makers and stakeholders need data, and you need data to make good decisions in operations.

Using tools to combine existing data into dashboards and readable metrics is the best and easiest approach. Here, it's important to define or develop a platform that pulls data from all your sources, where everyone can access data in one place.

You can choose to collect data manually and ask developers and team leaders to move their data into the platform manually. For example, if you want a Quality Assurance team to share the number of bugs they find per week, it is a good idea for them to be familiar with that figure and forcing manual data sharing may seem like an ideal way to achieve that. However, manually adding data to a platform will add work for your teams and complexity for you, because you will have to account for manual error, and you will need backups to keep data up to date. Manually adding data to a platform often results in mechanical copy-paste behavior, reducing the desired awareness factor.

In my opinion, automating data collection is a much better alternative to manual sharing. There are numerous tools you can use to collect data from across tools to automatically integrate it into your platform. The advantage is that your data will always stay up to date and it will always be there. You also won't have to worry about stressing teams with unnecessary work.

At the same time, you need to define processes so that relevant people remain aware of data. Automation can make data less actionable, shareable, or communicable because no one is handling it directly. You can account for this by using data connectors to display information in dashboards and data visualization apps which can be shared to relevant teams and stakeholders. You can also create views in different apps or create links to a wiki, intranet, or your documentation platform.

From there, you can work to create a culture of data review, where teams regularly analyze and act on data. Here, you can create a weekly, bi-weekly, or monthly meeting focused around reviewing data, so that teams make it a habit to review and consider metrics and information as part of their process.

At Nmbrs, it was important that we be able to automate data collection. At the same time, we wanted to ensure that chapters were aware of data and KPIs, because those chapters are responsible for quality. Our solution was to create a monthly meeting with chapters, where we organize data around monthly dashboards which are used to drive actions and improvement for the next period.

Using data in this way keeps relevant people informed and enables stakeholders to use data to take actions, set targets, and improve or decrease certain KPIs. Our cycle also helps teams to understand KPIs they can drive their own actions in ways that helps them to measure and improve. It doesn't matter if targets are delivery rates, fewer bugs, automated test coverage, etc. so long as teams can see how their actions contribute to improved metrics.

While KPIs and data are important, they aren't everything. Always validate data with the individuals doing the work to ensure that what you are measuring reflects reality.

Strategy Tip: Keep teams involved with KPI measurement so that it stays relevant and actionable rather than remaining "just" data.

Engaging teams with data

Data is easy to collect and most organizations now benefit from an overabundance of it. Using data is much more difficult, because once you put effort into collecting and organizing data, you still have to get people on the work floor to actually use it.

Driving a data-oriented culture often means pushing a cultural change across the organization, because most people don't naturally orient towards using data. My recommendation is to build a Data Guild, linking specialists together, to promote using data, integrating weekly dashboards, infographics, and charts to drive engagement and interest.

At Nmbrs, we use guilds to share data across multiple squads. This allows anyone who likes data to be involved, so data spreads throughout the organization, rather than being trapped in a silo.

It's also important to involve teams in using data metrics. For example, as previously discussed, you can create weekly or monthly sessions where teams get together to review data dashboards and make performance or strategy updates accordingly. You can also integrate triggers, such as when support tickets reach a certain threshold, it starts a support escalation process.

Here, teams are often unaware of what data they need or what can be valuable. Challenging teams to consider their own needs, determine which KPIs are relevant to them, and helping them to build structures to collect that data is one option to drive engagement. Plus, actively defining KPIs with teams will help them to remain engaged. If they participate in defining what's being measured, they are much more likely to actually use it, versus simply being shown "meaningless" numbers on a dashboard. Teams should understand how data will help them improve performance because it will affect how they approach it.

Even the process of sharing important data with teams on a regular basis will help to keep them engaged and interested. When teams know what their actions are achieving and how their work reflects in metrics, they can better decide what to do and when to steer the organization in the right direction.

Once you have established a culture of data-usage, it will likely continue. As the tech COO, you will be able to easily monitor your operations and value streams by looking at data dashboards, which you can fine-tune to look at teams, processes, bottlenecks, and other root causes of inefficiencies using data.

Strategy Tip: Involve teams with data to ensure adoption and integration.

Conclusion

Data and information encompass the full journey from vision to value. Data helps you measure how products are running, how value is perceived by the end-user, drive feature development based on end-user perception, and define performance within teams and the organization. Data ties into every other pillar of the organizational model, and every aspect of your organization.

Developing a data and information strategy that defines not just which metrics you collect but also how to use them and how to integrate them into teams is crucial.

PART III

STRUCTURING
TECH OPERATIONS

The Vision to Value Framework aims to provide a structure and framework to support operations. In Part III, I share practical application of the concepts and ideas discussed in this book, to offer more actionable information.

Applying concepts to a live environment is often difficult because every environment is different, and everyone handles product development in different ways. At the same time, I believe there are common considerations, no matter what or how you are implementing.

Product Development

"If you are not embarrassed by the first version of your product, you've launched too late."

– Reid Garrett Hoffman, Entrepreneur

Technical operations are traditionally intrinsically tied to product development. Operations must source materials, workers, and processes – importing raw materials which leave as viable and saleable products. In tech companies, this process is even more interconnected.

Your teams are the most important resource in your development and deployment pipeline, because their work is the raw materials going into your product. Providing them with the structure, processes, and procedures to maintain the quality of products and services already released is crucial to your success as an organization.

Setting up the proper structure and processes will enable you to react quickly to events inside a working environment, so you can maintain quality where users see the product. Here, escalations, 24/7 monitoring, DevOps, and other methodologies are extremely useful.

Your operations must be structured around creating an efficient deployment channel, enabling product design and development to run smoothly, while supporting existing maintenance cycles inside the organization. Here, I will begin to cover some of the best practices and approaches for doing so.

Begin with clear requirements

If you were to ask the average developer what delays or impedes projects the most, they'd likely answer "bad code", "technical debt", "poor technical solutions", "etc.". However, in my experience, these problems are secondary to issues relating to poor communication during the project kickoff. Tech teams often don't have a clear understanding of deliverables, who they are for, why they are being developed, or what (business) problem they are solving. Instead, teams receive communication in channels they don't understand, with unclear requirements, or other issues. Guided by misinformation, they begin working in the wrong direction.

Clear communication during kickoff is crucial to product development. Clearly communicating expectations, requirements, and the root problem or need behind a product is essential to developing something that offers value. If teams don't have proper or clear information, they cannot build a product to meet a need or solve a problem. Stakeholders must have the channels to clearly communicate what they want or need to developers.

As Tech COO, it's crucial that you create processes to push required information to teams. This also includes developing channels for communication, defining requirements, and standardizing everything across teams.

It's true that there are many methodologies but it's crucial that everyone is on the same page, because if every team communicates in their own way, pushing information is no longer scalable.

My advice is to create one standard channel for pushing information (such as storyboarding) and combining it with kickoffs, a requirements analysis, and research. If you begin every project with a kickoff integrating the full team and any stakeholders, who each offer their own input, teams will have a much better basis with which to begin development.

In these sessions, stakeholders should discuss details and request estimations. What are the project requirements? How do they fit into company vision? What is the reasoning behind the project requirements? How is it contributing value? What should development time look like? When you are not able to estimate development time, teams will start working without clear scope or boundaries, which will always result in delays. Only when you have clear time estimations will you be able to allocate work items to sprints and to predict delivery-time.

Teams should understand the problem, why the problem exists, and should be able to invest as much time and effort into the kickoff as possible so that the project begins on the right track. Why? It's easy to attempt to fit a kickoff into a one-hour session, but you often cannot communicate everything in just one hour. Teams may think they grasp the problem and then go on to develop the wrong solution. Investing more time into creating a viable and valid solution will save time and money in the long run. If everyone agrees on what is needed, why it's needed, and what the finished product will look like, the product your team is working on is that much more likely to meet parameters and deliver value.

Creating clear channels for communicating project needs also means choosing a methodology with which to do so. SaaS companies often develop new things, which requires more creativity and more Agile approaches to standard project management. Ensuring individuals have a strong knowledge of the end-goal, the problem they are solving, and the customer will prevent your project from becoming chaotic. You need Agile to achieve this.

Agile processes and operations create numerous methods to push information to teams. User story mapping is one such method. Here, user story mapping allows you to describe a persona, defining what the user wants to do and why. Understanding the "Why" of what the user wants to do gives teams much more insight into what will fix the problem, so they can take a proactive approach rather than simply building something for the sake of building it. Story mapping is just one channel you can choose to use.

At Nmbrs, we host kickoff sessions each time we start a company goal. These sessions start with the why of a goal, explaining vision and connection to the business plan. Then, we move on to "what" or the scope of the goal. This information allows the Product Owner and teams to start their own "slice and dice" sessions, where they break the solution into epics and user stories. They also create a high-level estimation and prioritize major milestones around user value. Finally, the resulting plan is discussed with teams including stakeholders, product owner, and development teams and updated according to their feedback. With this approach, we were better-able to predict delivery. It also empowered teams, who could use background information as a basis to design their own solutions.

Strategy Tip: Investing more time into the kickoff phase empowers teams while making delivery much more predictable.

The Development Pipeline

The development environment is the core of any SaaS organizing, comprising the only place where value is produced in the form of a viable and saleable product. Developing an infrastructure supporting a scalable and solid development environment is crucial to your ability to scale and to continue providing value to the customer.

In the development pipeline, you build, deploy, test, and release products, features, and processes. Each of these steps must be handled seamlessly, meeting deadlines, quality expectations, and integrating into a live software environment with minimal disruption. Implementing effective best practices to make software development fast, reliable, and repeatable is key to a sustainable development environment.

Achieving these best practices means creating individual environments for development, quality assurance, testing, and production. It also means defining clear ownership and access rights for each environment. Creating a defined development process where individual role responsibility is clearly delineated at each stage and in each environment will allow you to control each step. Here, you should pay attention to local development, team testing, global testing, and the production environment.

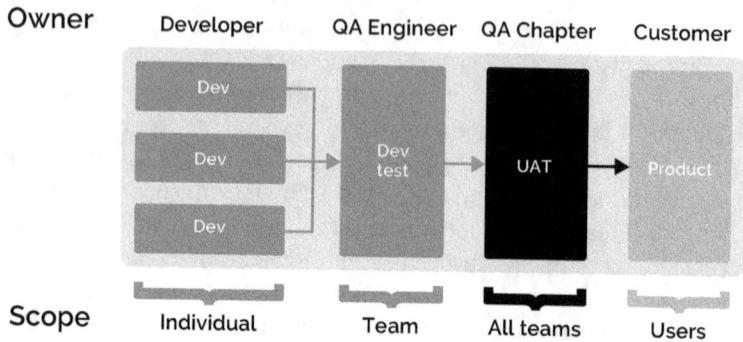

Fig.33. Development pipeline with ownership

Local development: Local development encompasses the developer's local machine. This is where development assets and code editors are stored. It's also where the developer writes code, debugs applications, and runs the application locally. Ideally, the developer should be able to run the entire application on their own machine. Unfortunately, this isn't always possible, especially for larger applications. If not, your development environment still has to run automated testing from other teams, so that developers can validate their changes before pushing code to the next stage of pipeline. Developers must be able to assess the impact of their changes and creating a local development environment capable of performing and accessing unit testing (at a minimum) is crucial to this. This environment encompasses the individual scope of the developers work and should belong to the developer.

Team testing: Team testing is a test environment, where teams integrate work from individual developers and thoroughly validate the impact of changes against the entire application. Quality Assurance engineers from each team should own this environment. QA should deploy new changes and should restrict developer access. The more restricted the access, the more developers are forced to use the deployment pipeline and scripts to make changes. Doing so is a much better approach than giving developers full access to the team testing environment, because changes are visible, repeatable, and traceable. It's important that developers not be able to make changes to the testing environment because QA needs to trust their testing environment, so they can trust results.

Global testing: Global testing is a shared environment where you can integrate work from different teams and run complete integration testing. This is ideal for several teams working around the same code base. This environment typically resembles a production environment, making it an ideal place to validate non-functional requirements, such as performance and security. Global testing impacts all teams and should belong to the entire QA chapter.

Production: This is the environment you provide to end-users. The process of pushing changes from the final testing environment to the production environment should be automated, so that fully validated changes are automatically deployed to customers.

Each time you push changes to the next stage of the development pipeline, it is a form of validation, which gives you more assurance the "real" update will succeed.

Here, it's also crucial that you create an environment where developers have to naturally follow the steps in your formal development process. When everyone follows the same steps and processes, everyone can see what has to be done and where. This helps to create accountability and ensures everyone knows who is responsible at any given stage.

At the same time, each environment must have a clearly defined owner. The first owner is logically the developer, who must own his or her own development environment. QA must own developer testing in the scope of the team. In the production stage, the user, who is the most important stakeholder, becomes the owner.

Why is it important for QA to own their testing environment? We discussed earlier that a QA must be able to trust their testing environment so they can differentiate between product or code defects and infrastructural issues. For example, if the QA were to test a feature and discover that something isn't working, they would likely contact the developer about it. The developer would then most likely connect to the team test environment and their source code to debug in that environment. The result is very often that the developer finds something strange or out of place in the testing environment and chooses to manually fix it there. Then, the feature obviously works.

The problem is that even if the source code requires fixing or adjustment, it must be done in the live production environment, and not in testing. If the product works in testing after manual changes to the source code, there is a high risk of the product not working in the live environment.

Creating a process where QA pushes code back to the developer to fix it in their environment will slow progress, but it ensures code will work in the live environment. Manual intervention is a quick fix, but it only adds risk at a later stage.

Strategy Tip: Map the development pipeline to separate environments with clear ownership and purpose for each.

Bootstrap from Source Control

Maintaining quality and consistency in a development structure is crucial. My recommendation is to integrate everything, including new source code, data schema and configuration information into source control. Why? Tying everything to source control allows you to easily restore a working version of software whenever anything goes wrong. It also ensures that your development structure is easier to scale with engineers because new build environments can simply be bootstrapped from source control. This allows you to better maintain quality.

Any software application can be split into source code, assets, data, and configuration:

Source code: The application source code in the form of computer instructions in one or several computer languages such as C#, Java, Python, JavaScript, SQL, etc.

Static assets: All static content such as images, style sheets, icons, etc.

Data: Includes data in databases which is needed to run your application. This is not about user data, but system data that your product needs to run, like navigational structure, user roles or feature permissions.

Configuration: All settings connecting your application to its environment, including database connections, file paths, server configurations, access control, etc.

Incorporating everything into source control is advantageous for several reasons. For example, if you have everything in one place, you can automate deployments. You'll be able to more easily roll-back changes, restore previous iterations, and create patches for new documents. Onboarding new developers also becomes simpler, because you can auto-configure new build-environments.

Bootstrapping everything from source control facilitates test-environment setup. Manually setting up cases is one of the most difficult aspects of developing a proper test environment in sync with developer pipelines. If data isn't in source control, manual is the only option. Maintaining everything in source control also allows you to more easily run test cases in different environments – such as the test server of a specific team – because you can manually replicate all tests. With everything in one place, testing becomes more easily replicable and automatable.

Pulling data, configuration information, and code also helps you connect source control to project development. Once you've created a strong link, you can develop transparency in the form of traceability and change records, so that every update is traceable to an event, developer, or team.

At Nmbrs, we integrated this strategy to great success. Our first step was to integrate all artifacts including code, database, configurations, libraries, tests, and other resources into the source control system. This enables us to bootstrap the full application from a specific version without the need to pull from external configurations or database states.

Setting up this process will help you connect source code to vision, even when it's to a ticket – because even tickets link to initiatives, epics, business plans, and therefore, to organizational vision. This will, in turn, help you track code changes and categorize them as new product development or maintenance. You can then easily trace whether code is moving the organization towards goals.

> **Strategy Tip:** Bootstrap all development assets from source control.

When features don't fit in a sprint

Branching is a very commonly used development strategy. Here, you organize code changes into different branches such as a development branch, a test branch, and so on. Once the code is ready, you move it to a live branch and push it to customers. Branches allow you to retain different versions of code, so that developers can push new changes to test and test can push "finalized" code to a live branch.

The most common approach to branching is to map branches with the development stages from your pipeline, typically, development, test, and production. Agile development processes typically divide work in sprints of 1-2 weeks. Developers push code to a development branch during the sprint. Later, QA engineers merge this branch to test, which is then deployed to the production environment. It's also very common to create a new branch for each sprint, so that you know what is completed on each sprint. However, this process becomes problematic when you begin to develop larger and more complex products which don't fit onto one branch.

Agile coaches will always try to break work into smaller user stories, but code doesn't always cooperate. If code is too messy and too risky to commit to source control, you can create a code branch for the feature or solution instead. This gives teams the freedom to continue working on that feature without time pressure.

Here, the theory involves creating a code branch where teams to merge updates. Sprints are merged into this branch as they are completed. When the full project is completed, everything is merged into the development branch in one go.

However, this can be problematic. For example, code branching typically results in having two separate copies of your application. In most cases, the original development branch will have changed. Maintenance will keep running, other teams will continue merging their branches, and so on. If you attempt to merge everything at once, there will be conflicts and the process of merging may be very complex.

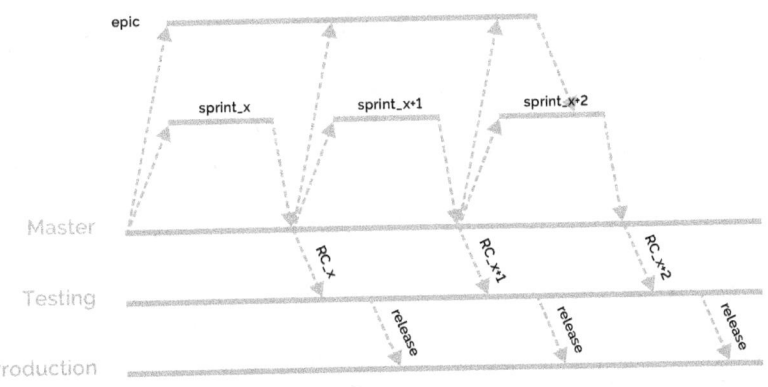

Fig.34. Sprint branching strategy

For this reason, and others, many developers avoid branching when possible. "Feature Flags" are one technique that can help you do so. Here, you develop new features on a "master" code branch. The Master code branch is merged with the live branch as new features are finished. The Master also takes the place of individual dev branches by ensuring features are merged as they are created to prevent outdated code or merging issues. How can you merge code to the production environment before every feature-in-progress is completed? Feature Flags or Toggles simply prevent unfinished code from being used.

Here, incomplete features are switched off in the production environment, meaning they are invisible to front-end users. If a feature is bug-ridden, the new code impacts other features, or otherwise isn't ready, it remains switched off.

Working on a single master-branch means that code never runs the risk of becoming outdated. When there is a merge conflict, you can detect it right away and it can be fixed with minimal effort. This also allows teams to develop entire features which are pushed to the production environment before they are finished, but which are invisible to end-users. While not always crucial, this is important when two new features interact, but one is ready to go live and the other isn't.

Adopting this strategy enables you to foster continuous integration and faster development because developers have access to all code.

My recommendation is to avoid branching whenever possible because merge conflicts are complex, require manual intervention, and can be difficult to spot until something breaks. Branching too much will create delays and added costs, because you will have to rework outdated code and will have a long and tedious merging processes.

> **Strategy Tip:** Keep branching to a minimum by using feature flags.

Backwards compatibility development

Improving software often means changing how code works, how it connects to other software, and how it integrates different components. Making these changes can be risky in that it may prevent necessary features or connections from working. Changes may cause dependencies and delays because teams have to wait before they can develop a connection or integration that works. If you have to update every component of your software to meet new integration standards and those components are owned by different teams, or worse, third-party vendors, you need a considerable amount of coordination to prevent costly delays. You'll have to update all systems before rolling out changes, which will result in delays.

Developing backwards compatibility helps you to avoid this issue because you can ensure all new connections work with the existing ones. This gives teams and external component vendors time to update their versions, while giving your team the option to push their new connection to the production environment.

Doing so typically requires a technique known as Version Control, which is common with API changes. If you're changing operation parameters, creating complex structural changes, or removing or renaming an operation, you'll likely want to create a new version of your API. This indicates change in either consumption requirements or the resources offered by the API. Here, it is crucial to document changes and create an easy-to-follow process for other or external teams to adapt their own products and features.

Versioning is especially complex when you consider other teams and how they connect their components. For example, your database is likely a shared resource with numerous components connected to it. Simply renaming a database column would likely break several components. Instead, you'd want to use a backwards compatibility technique, such as creating a new database table that functions alongside the existing one.

If you can create a new version of a product with dependencies and leave the old one in place, teams can adapt and integrate in their own time. This ensures continued backwards compatibility while giving developers time to update their own products, without creating dependencies. It will be extra work, but it will also help you to deliver faster. It's also extremely relevant when integrating external components from third-party developers, because they will never be ready for change.

Maintaining multiple versions of a product will become resource-intensive, so any instance of duplicate versioning should be temporary. Create a process where you clean up duplicates after a defined period to ensure long-term stability and semanticity.

Backwards compatibility techniques allow you to deliver more quickly by avoiding dependencies and external dependencies. Integrating them into your processes is important if you want to ensure seamless changes.

Strategy Tip: Incorporate backwards compatibility techniques when developing software.

Deployment pipeline

The deployment pipeline is a series of processes that push software from version control to the customer. The deployment pipeline is the mechanism that delivers value, while testing for quality and performance. It is one of the most important processes in your organization, because it is where you will take the step of delivering value to the customer.

A structured deployment pipeline shifts focus away from (solely) development processes and towards the quality of the delivered code. Teams very often focus on releasing as many new features as allowed by their scrum processes and sprint cycles, rather than allocating an equal amount of time to deployment and delivery. Integrating a deployment pipeline to shift some attention away from development and towards quality control will result in higher customer satisfaction, because it ensures code is validated and tested before it reaches a live environment.

The deployment pipeline is designed to avoid or prevent waste during development. It is often used to provide feedback and validation throughout development. It also enables you to automatically fail code that doesn't pass early stage testing, so you spend less time and money producing something that doesn't work at a more complex stage.

Achieving this "automatic" testing involves creating a series of stages in your deployment pipeline. Here, you separate simple tasks from those that take more time and conduct value stream mapping for the deployment workflow. This can then be used to help non-technical team members understand where different aspects of deployment add value.

Your teams should be involved in creating deployment pipeline stages. Almost everyone implements deployment pipeline stages including "build" "testing" "staging" and "production".

In the build stage, software is compiled and packaged, Here, input is typically handled through source-code. Build-stage output can be automated and tested based on changes to the source-code repository, enabling immediate feasibility and quality testing.

During staging, the build is moved into a staging environment – typically a clone of the production environment – where testing verifies the functionality and integration of the new build and its capabilities. Good automation should also test for regression to ensure the new version doesn't break old capabilities. Testing should also automatically check code against quality and performance standards at each stage of development.

Production is the final stage of most deployment pipelines, where software is installed into the production environment. In a SaaS company, this must be deployed while the product is running – maintaining full software usability for the end-user during the integration. Here, you must test to ensure that the new version is working as expected.

In a simple pipeline, your stages will run sequentially, moving from one to the next and pausing when errors occur. In more complex pipelines, you may have multiple instances of several stages, allowing parallel development. In most environments, the early stages are simple, fast, and designed to provide rapid testing and feedback to teams for proof of concept, source-code validation, etc. As development progresses, stages become increasingly more complex when you move software into a production-like environment.

Creating a deployment pipeline also allows you to introduce automation, which will enable scalable continuous delivery. Continuous delivery –where you break large release cycles down into smaller ones and push them to customers more quickly – reduces risk, improves quality, and supports steady scalability of the application. However, it is too costly and time-consuming to maintain as you grow without automation. Automating processes allows you to increase workload and releases without adding extra overhead and while ensuring that quality remains consistent with each product release.

Automating your deployment pipeline also helps you auto-test for defined quality parameters at each stage of development. Each time developers submit code to source control; your pipeline should begin an automated review. Here, you can measure code quality using metrics such as cyclomatic complexity, lines of code per unit, or adherence to guidelines.

This helps ensure consistency across code submitted by different developers and teams. From there, you can automate unit testing and integration testing so teams can easily monitor if components work together. This is important for complex applications developed by several teams or individuals working on their own.

Automating these processes reduces bottlenecks, because testing and approval just happen. Teams can simply define processes with set delivery stages, where code is tested, aligned with the software, and validated. New products are then tested at each successive stage, resulting in an end-deliverable that has been tested and validated in a working environment. If something does go wrong, the "problem" can simply be pulled back to source control and re-released with the next update.

Automating your pipeline reduces potentially error-prone manual processes, speeds up flow between processes, and enables seamless continuous deployment. It also enables transparency in the pipeline, generating data at every stage, so you can see how tools are working, track the health of running applications, and trace code as it moves from production to deployment. This will aid in auditing and quality assurance, so you can quickly identify and fix issues when they occur.

Automating testing environments enables you to automatically run testing from different perspectives. Automatically provisioning, deploying, setting up, and configuring testing environments will greatly improve the speed and efficiency of the process. While it isn't always possible or desirable to create a testing and validation environment that functions entirely without human involvement, it can be massively beneficial to automate most or all testing processes. For example, automating the deployment pipeline removes the need for manual-error prone tasks, helps teams to identify flaws early on, and ensures work is visible from the start. This will, in turn, make onboarding and cross-team collaboration that much easier, because everyone will know what's going on and where.

Your approach to setting up deployment automation should differ depending on whether you're building a development pipeline from scratch or are implementing automation into an existing project. If you're building a new pipeline, your primary goal should be to start a test project and use it to identify and implement the stages requiring automation.

If you're integrating automation into an existing pipeline, you'll have to take a different approach. Here, you should step back, review the pipeline, and identify the largest bottlenecks. You may also already have some automation in place, which could be supplemented and improved.

Most organizations implement tools in combination with an orchestration framework to tie them together. For example, a good automation chain might include source-code management tools such as Git, build tools like Make and Maven, a CI server such as Jenkins, configuration management (Ansible or SaltStack), Deployment tools like IBM® UrbanCode® Deploy, and actual testing frameworks such as xUnit or Selenium.

As with any tools, automation tools are a means to an end. There are no best tools, only the best tools for what you are doing. Anything you implement should include a continuous process for maintenance, reviewing quality, and reviewing efficiency to ensure that you are using the right tool. Teams must be able to trust their tools.

Depending on the size of your organization, you may create an automation framework and have everyone maintain it, or you may be able to create a dedicated team to maintain and improve the framework. As you grow and automation takes on more and more labor, creating a dedicated team will become more and more important.

Strategy Tip: Fully automate your deployment pipeline.

Quality Gating

"Make It Work, Make It Right, Make It Fast"
—Kent Beck

Quality management is one of the primary advantages of a deployment pipeline, because the deployment pipeline allows you to integrate automated quality gates. Here, each gate is an automated validation process, where code must meet criterion for code analysis, performance, security, or functionality. If code does not pass at any stage, it does not go on to the next stage of development.

This process of quality gating screens easy-to-test-for problems early in the development process, with new gates for each subsequently more-difficult-quality-test. This ensures that no code with simple-to-screen-for errors makes it through a more costly or time-consuming validation process. If it doesn't work right away, you don't invest more testing into it until the "easy-to-test-for" problems are fixed.

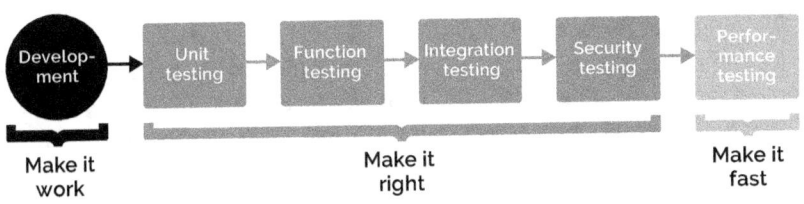

Fig.35. Quality gates in delivery pipeline

Most deployment pipelines divide quality testing into 3-5 gates including code analysis, functional testing, integration, security, and performance. Here, your first gate should be code analysis to test micro-level quality and to run unit testing. Your second gate should validate that the module functionally works. Your third gate should validate if different modules work together according to the functional scope. The fourth gate should test security and validate items such as unauthorized access prevention or cross-site scripting or forgery requests. Finally, the fifth gate should validate load testing.

Gating order is almost as important as establishing gates. If you prioritize screening based on the speed and investment required for each test, you ensure no code moves into a difficult testing stage with simple errors. This will reduce waste. For example, a unit test can be run in seconds or minutes so it's easy to run and validate. Later tests require significantly more time investment and should be run after validating content for simple tests.

Quality gating can help you to improve the quality of software, but it also helps in other ways. Test visibility – where you know how many tests are performed at each gate and validate that these tests are performed on every module – ensures you can see where things go wrong. If you have functional problems, you can assess functional testing coverage. If you see security problems, you can assess security testing. This allows you to directly map gates and existing testing to symptoms and quality factors. You can also use this same increase in transparency and visibility for compliance.

Enabling this type of gated deployment pipeline often means breaking releases into smaller, focused updates. Breaking major updates into cycles of smaller recurring updates enables you to integrate routine checks, which are more likely to catch small issues before they become a problem. It also helps developers trace problems rather than sifting through multiple small updates inside a batch update.

Smaller updates reduce the risk of high-stakes failure, because implementation and code release are naturally smaller. Smaller portions of code are validated and tested individually, giving teams more room to understand where problems are and how to fix them.

Here, you should keep both functional and nonfunctional quality in mind. Many QA engineers test for functional quality (what the product does) but leave out non-functional aspects of quality because they are less obviously relevant.

However, customers are also concerned with non-functional aspects such as performance, app speed, security, etc., each of which can be just as annoying as problems with functional quality. For example, an app that takes too long to load is almost as useless as an app that doesn't load at all. Integrating automated quality controls as part of your deployment pipeline can help you to avoid this common QA blind spot.

Developing well-structured quality gates makes it easier for teams to update someone else's modules or source code because they can run required quality tests on their own, without the need to ask for help from another QA or tester.

Quality gating also (logically) helps you improve quality control. Having the ability to measure different aspects of quality independently, such as by code coverage, page load, capacity, etc., means you know where to focus efforts and improve quality.

> **Strategy Tip:** Integrate complex functional and nonfunctional testing into your deployment pipeline.

Release frequently

Product releases are crucial to any organization. When tech giants like Apple or Samsung announce the release of a new product, the world gathers round to listen. At the moment of release, your organizational vision is transformed into real and tangible value for the customer and end-user. It's only natural that you'd want to release a perfect product, delivering the most possible value to the customer. However, the process of product releases should be intrinsically different for a SaaS company. Where a tech giant like Samsung must finish their product, often building on a previous one to create something complete and whole, SaaS companies should often focus on fast product development and release schedules, delivering value more often, rather than a complete or "perfect" product.

Tech giants like Facebook already embrace this tactic wholesale, with a culture of continuous delivery dating back as far as August of 2013[19]. Here, Facebook rolls out code twice per day, including a standard daily push managed by the release engineering team, supplemented by a secondary team in New York. This twice-daily push enables Facebook to scale small releases to their team of 1000+ software developers and engineers, creating incremental pushes that reduce the risks of pushing large amounts of code to users all at once.

This process of constantly pushing content to Facebook enables the teams to push a large number of changes to billions of users, with as little disruption as possible. Facebook integrates continuous development with a great deal of automation to prevent bottlenecks, reduce risk to the end-user, and to manage the impact of new releases by making them smaller.

---

19 https://www.facebook.com/notes/facebook-engineering/ship-early-and-ship-twice-as-often/
10150985860363920/

Continuous delivery shifts attention away from finalizing and releasing a perfect product and towards creating a minimum viable product, which is then developed and added to in short cycles, so that it can be reliably released at any time. This approach reduces costs, cutting the need to invest an entire budget and years of work into a product before delivering change, enables incremental updates and frequent new features so that end-users continue to see value, and enables organizations to deliver software with greater speed.

When each release brings value, the end-user continues to see value and therefore gains incrementally more from each update. Similarly, frequent product releases increase validation touchpoints, giving you more opportunity to validate new products and features, more opportunity to review and change those products to something that best-benefits the end-user, and to minimize the risk of change.

The more often you release products, the more smoothly it will go. However, creating a process of continuous delivery also shifts focus away from implementing massive changes overnight to introducing small and well-tested changes. This reduces the likely number of bugs and integration problems you would see with larger or more comprehensive updates.

At the same time, no testing measures will completely prevent bugs. If you're pushing software to 3,000 users and the 1 in 1,000 chance of breaking your code results in an outage for 3,000 users, that's a lot of lost value. If you have 10,000, it's an even larger problem. The bigger your user-base becomes, the riskier pushing updates to everyone becomes.

Here, you can integrate standard risk-reduction strategies to prevent major update problems; segmenting, beta testing, and parallel environments.

Segmenting releases is the process of pushing releases to small groups, validating that release, and continuing to push to more groups. This software rollout allows you to test usability, functional and nonfunctional quality, user satisfaction, and the deployment process so you can tweak, optimize, and improve as you go. If something breaks, it affects a small number of users, rather than causing an application-wide outage. At the same time, you can improve software and add more value to the production chain.

Facebook implemented a similar strategy in 2016[20], where they push updates to Facebook employees, then, if all goes well, push to 2% of the production system. Once validated, Facebook rolls the release out to all users, and continues to validate throughout the process. This system replaced their previous "three-push-a-day", with the notable result of eliminating the need for hotfixes in the case of sudden problems during the rollout. Instead, problematic code is simply committed to source control and replaced in the next release. Avoiding single push rollouts also reduces the impact for users because problems are more likely to be diagnosed before rollout. Finally, continuous delivery forces teams

---

20 https://code.fb.com/web/rapid-release-at-massive-scale/

to continue to adapt and adopt new tools and functions so they can continue to meet the needs of seamless delivery.

Beta features give a small group of individuals the opportunity to request changes and new versions. Beta programs can help you diagnose bugs and potential problems with your most active user-base, garnering live feedback and real-world use-cases in an environment where users often expect software to break.

Creating a parallel production environment will enable zero downtime updates, for continuous delivery while reducing risk in the live product.

Here, tactics like "blue-green", first described in the book "Continuous Delivery" by Jezz Humbel and David Farley[21], are used to prevent issues. Netflix famously uses a similar idea, which they label "red-black"[22]. Their concept is to create a clone of the production software. New products and features are integrated into the red-black, or the environment not currently in use enabling a full operational testing without affecting the user-environment. Once validated, switching over to the new version is as simple as changing router configuration to the "red" version. If something goes wrong, you simply switch back to the black environment to restore the pre-update product.

Blue-green deployment is similar, except that in a blue-green environment, both servers may technically receive requests at the same time.

Both strategies enable pushing software releases and updates across a live application without risking a total crash or outage. Pushing software to a small number of users, validating, and then rolling out to a larger user-base reduces risk, allows you to catch initial errors quickly without affecting everyone, and improves total user satisfaction – adding to total value.

In some cases, you won't be able to divide a project into shorter delivery cycles. When this happens, it's also a good practice to include toggling, which we discussed earlier. Here, unfinished features are included in the deployment package, with a flag or toggle to hide them from the end-user, providing there is neither functional nor technical risk. This will prevent problems with overhead branching relating to code merge and synchronization.

In the early days of Nmbrs, we had a very small development team. We were able to push updates several times per day, which was possible because of the small size of the team and product. However, each update impacted users, because they were kicked out of the application and would have to log in again. As a result, it was quite disruptive. We maintained it for some time because it enabled us to release improvements and bug fixes very quickly. Eventually we organized a weekly deployment cycle where we would release a new update each Thursday evening. Why Thursday? If something went terribly wrong, we could fix it on Friday, which was normally a quieter day in terms of usage. Some

21 http://www.synchronit.com/downloads/Continuous%20Delivery%20-%20Reliable%20Software%20Releases%20Through%20Build,%20Test%20And%20Deployment%20Automation.pdf
22 https://medium.com/netflix-techblog/deploying-the-netflix-api-79b6176cc3f0

Fridays ended up including a great deal of bug fixes and patches. Customers asked "Why update every week? Why not once per month?" We had to explain that if we were to release everything at once, all the issues would hit all at once. We'd need an entire week to solve all the problems released in a month of development. Instead, we started pushing smaller and more frequent updates to minimize the risk and disruption for each update. We introduced micro-services, speeding up the update cycles for several teams, so that they could adopt a process of continuous delivery. The result is that our release cycles are much faster and more convenient, resulting in reduced impact for the end-user.

Strategy Tip: Release your product as frequently as possible.

Continuous Deployment

A fully automated deployment pipeline enables a process known as continuous deployment, where software is continuously released in short cycles through automation rather than manually at the end. Here, your base processes should involve structuring build-servers to receive source control automatically so that all new updates go through the build server and source control, rather than being individually pushed to deployment by developers.

CONTINOUS INTEGRATION

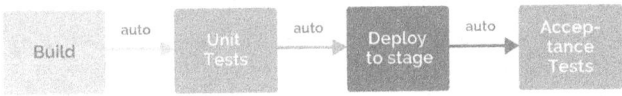

CONTINOUS DELIVERY

CONTINOUS DEPLOYMENT

Fig.36. From continuous integration to continuous deployment

Setting up continuous deployment also requires continuous integration, which is a standard for many SaaS companies. Here, you continuously integrate new changes into version control, automatically move changes into the testing environment, and then automatically deploy changes once they pass manual acceptance in the Test Environment. Making this shift allows you to deliver value much more quickly, simply because small and incremental changes will be processed and pushed to the production environment immediately.

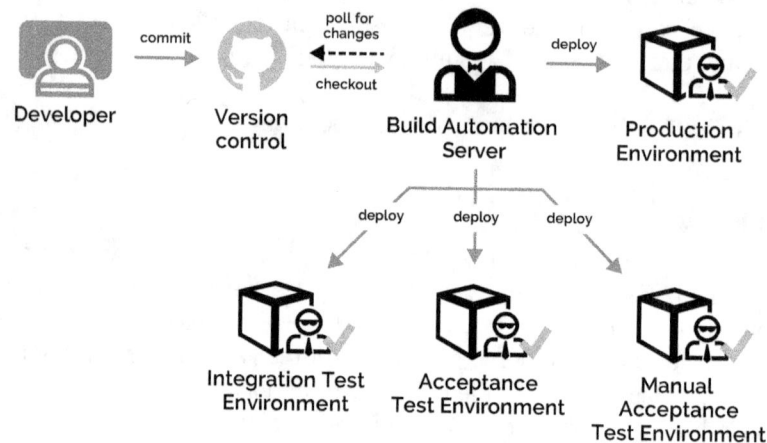

Fig.37. Integrating and automating delivery pipelines

In continuous deployment, it's common practice to have developers pull from a source control repository or library to begin work, push their builds to source control, and have those builds pushed from source control to deployment. All updates go through the same source, are logged and recorded in the build server, and can be removed if something goes wrong.

This will, in turn, create more value through earlier return on investment and smaller upfront investments required, earlier feedback from users, and faster iterations of improvement.

At Nmbrs, we implemented continuous integration with a build server as the core tool of our deployment pipeline. The pipeline includes several stages, which measure code metrics, perform unit testing, and run automated acceptance testing across several components. Higher level UI testing, where teams create their own test cases, ensures usability.

Deployment is fully automated, using database deployment scripts with database lifecycle management (DLM), and then deployed using a tool which fetches release candidate packages and deploys them to a target environment for integration, testing, staging, or production.

Each step of this process is fully monitored, with overviews of each running version, predefined workflows, and approval chains to ensure that every step is followed through.

Strategy Tip: Strive for full systems integration and automation to develop, test, and deliver software.

Early Validation

While late-stage validation is crucial to detecting bugs and ensuring seamless rollout to customers, it's crucial that you begin the process and code validation well in advance of deployment.

Validating and ensuring quality in anything you create or deliver is crucial to the short and long-term success of the organization. This applies whether you're rolling out a new framework, new code, a feature, or a completely new product. You must be able to verify quality and measure metrics as early as possible to validate success before investing a great deal of time and effort into it.

Here, your largest consideration should be that the later you discover a flaw or bug, the more it costs to fix. The further your code or feature moves along the development pipeline, the larger and more complex it becomes. Testing and validating early-stage code is fast and relatively easy, but becomes slower and more cumbersome as the code matures.

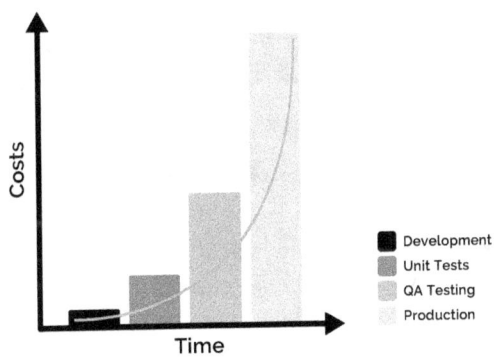

Fig.38. Costs of fixing per stage

Setting up processes for developers to validate and test as much as possible on their own machines is one easy way to solve this problem. If you invest in QA automation, enabling developers to pull tests from a library, validate their changes, and then push the code to the next part of the pipeline, you greatly reduce the likelihood of bugged or flawed code from moving forward in the production process.

Similarly, if developers are pushing changes into operating software, it's crucial that they validate the full solution before pushing changes. Here, "black/red" techniques work very well, because developers can fully validate solutions in a duplicate of the operating environment before they go live.

No matter what you're integrating, it's important to begin measuring quality and validating processes as early as possible. The further any product or framework is into the development pipeline, the longer it takes to fix.

Validating quality also allows you to recognize problems more quickly, so you can begin preventive measures to stop larger issues from cropping up. Prevention is the cheapest way to ensure quality. It's crucial to create clear code review processes to ensure developers are checking each other's code. Well-defined coding guidelines (documentation) and a defined code matrix are the two most important aspects of creating these processes.

ELIMINATE BUGS <u>EARLY</u>

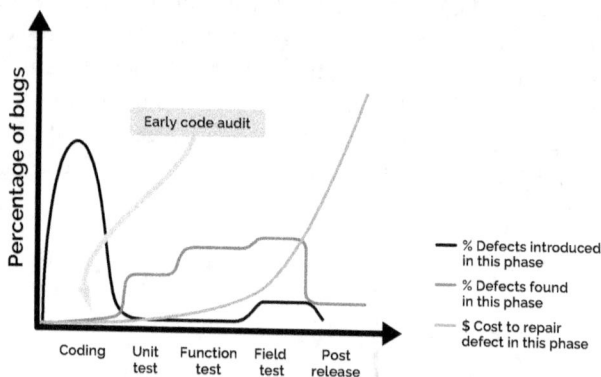

Fig.39. Cost benefits of eliminating bugs early

You must define concepts like "what is code quality?" "What is good?" "What is bad?" "Why?" In most cases, code can be measured against different qualities, such as security, maintainability, and complexity. If code achieves its purpose but is too complex to maintain properly, it's not good code. Here, you can use automation to check for basic quality standards such as cyclomatic complexity, lines of code per unit, or adherence to guidelines.

Most teams also want to check code based on Average Percentage of Faults Detected (APFD), fault severity, production incidents, quality over release life cycles, functional coverage, and so on. While there are dozens of quality metrics to consider, automated code analysis is not enough. You need process mechanisms in place to enforce or maintain code quality through unit testing, regular (manual and automated) code review, and ensuring code is testable.

Here, you should also implement code quality monitoring tooling into your development and deployment pipelines. Automatic tooling allows you to define metrics and baselines, so that when a dev writes a new code, it's automatically reviewed. If it doesn't meet standards, the tooling doesn't allow code to move to the next stage of the pipeline.

It's also important to implement processes where teams review each other's code regularly, such as through weekly sessions, to prevent issues. Peer review forces a consistent coding style across a project for better code maintainability, checks code against user needs and preferences which automation may not be able to test for, and gives developers space to share and develop ideas which could improve the quality of the project as a whole.

Completing this process as early as possible will ensure that code stays on track to meet customer needs from the start, so you save time and resources during development.

Strategy Tip: Integrate automated and manual code validation to begin measuring quality as early as possible.

Quality has a cost

Measuring development quality requires creating processes capable of tracking the cost versus quality or value of a product. This requires investment, but is crucial to maintaining the long-term sustainability and value of your development teams. If you cannot track the value of production versus what you are spending on it, you cannot properly allocate budget, cannot invest in flaw-prevention methods, and won't know how much you're spending on any given product.

If you measure the concept of quality versus value, you can create a more balanced view of quality. In this instance, the quality of a product can be measured as:

$$Q = \frac{F}{F + U}$$

Here, product quality can be measured as Quality = Flaws found by the development team(F) and then divide by the flaws found by users (U). If your team finds all the flaws or bugs and your users don't find any, it's high quality. If users find a lot, it's low quality. Here, you are measuring quality as user perception. If they don't see problems, you're a step ahead and they perceive your product as better quality. Investing more time into ensuring teams can find bugs before users do will improve this ratio. The cheapest way to balance this ratio is to create preventive measures, so teams find as many bugs as possible before they reach the customer.

Strategy Tip: Try to inspect code as much and as often as possible to find defects before customers do.

It's also important to measure software quality versus cost. Any increase in quality will require an investment. However, lack of quality also costs through lost users, poor user satisfaction, etc. You need balance to create a sustainable operation. Both no quality and too much quality are undesirable from a cost perspective, simply because no organization has unlimited resources.

You will eventually reach a point where your added value from quality investment is not as high or where it might not pay off. Here, your best option is to stop investing and focus on development speed, which adds more value than just quality.

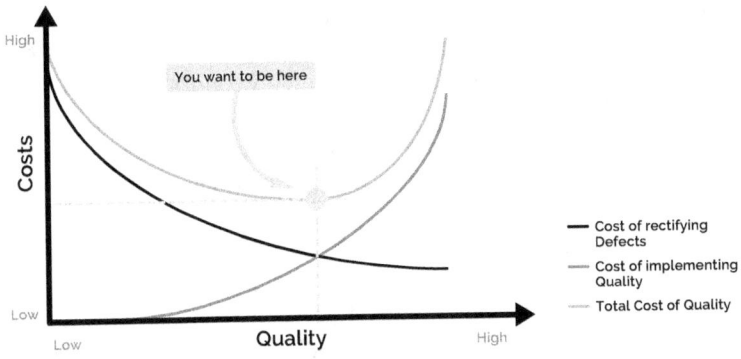

Fig.40. The costs of quality

Strategy Tip: Manage the value you receive from quality versus the cost of that quality.

Quality is about user experience

Think about a smartphone. Two brands like iPhone and LG typically function in the same way, to the same standards, and to the measurable degree. But most people would agree that the iPhone is better quality. Why? Quality is not just about basic functionality, it's also about usability, interface, durability, and other non-functional requirements. All this impacts the user experience.

Working to develop a culture of quality inside your organization ensures that people pay attention to more than just product functionality. Making the shift to provide consistency, usability, and a seamless customer experience will make the product better, but it will require everyone in the organization to align on this mindset. This extends to how and where quality assurance is handled.

Quality assurance most often ends after deployment. QA engineers put in work during the development cycle to ensure new issues will not be introduced into the production application and then quality assurance stops.

However, that update moment –when code is pushed into the production environment – is the moment when the most things can go wrong. Real users try the product, add in variables of their own data and how they use the product, and have the opportunity to break the product. Here, something that was high quality in a testing environment may fail to perform, it may be slow, it may not perform as users expect, and users may find bugs.

At the end of the day, users are your only meaningful quality filter. Their experience is the only one that matters. If they say the product is slow, it's slow. If they say it's buggy, it's buggy. If they say it's complex, it's complex.

The same applies to backend checks, where you check components during the deployment process. However, QA often don't take factors beyond those in the testing environment into account. Instead, most assume that if they check all the conditions and validate as much as they can, the product will work. The mindset is often that operations will ensure quality continues after deployment.

This is a problem because quality is an operational problem and QA teams have to embrace it. Automation is one solution. Setting up automated testing on live products will allow you to determine user-functionality in case the product breaks while live.

While it's valuable to have a strong QA solution in place, it's important not to rely on QA for providing quality. Instead, you have to develop the mindset that quality is everyone's problem. Teams should focus on every detail revolving around and delivering quality to the user. QA's job is to look for defects, but an absence of defects isn't "Quality". The tiny details and functionality are what make the product great.

> **Strategy Tip:** Ensure your QA team focus on the running product and use metrics that map user experience.

Development teams normally put a great deal of time and effort in QA to prevent new quality issues from appearing before launch. However, the quality of the product after launch is just-as-if-not-more important.

Good QA strategies mean recognizing this and implementing testing and triggers to identify quality in the running product from a user perspective. QA teams have to be able to measure and assess the quality of the running product. This assessment must be fed back to development teams, so they can improve the product.

This means establishing production quality baselines, such as page load time expectations, downtime percentage, processing time, etc. so that you have a basis with which to measure product quality. These metrics should relate to user experience, so that you can measure what users experience.

During quality assurance, you process metrics to ensure everything is working and then act on those metrics. However, development often uses triggers related to infrastructure rather than the user experience. Unfortunately, these triggers might not reflect how users see and use your product.

> **Strategy Tip:** Setup monitoring based on user experience metrics, not just technical triggers.

Front-end testing is another way to ensure the quality of user experience. Applications are typically composed of backend components powering the front end. If there's a defect affecting user-experience, it likely stems from backend components. When defects happen, developers often want to review backend components for the problem and then fix all possible components that affect the issue. Developers naturally want to make their products perfect.

However, this approach can leave gaps in user experience, because you'll always have components that aren't covered. The more complex your environment, the more difficult it is to trace components to the front-end features they effect. As a result, teams can spend a great deal of time improving backend components without improving user experience.

Validating if features are working from a user experience helps solve this problem. Testing from the front-end allows you to validate if things are working and then pinpoint which backend components could be contributing to the problem. If the feature works, the components powering it have to be working and you don't have to do anything else. This can be achieved by creating automated tests that simulate usage of features.

This will help you maintain quality in your running product, because you can recognize when things are going wrong for the user, not just backend components.

At Nmbrs, we experienced a great deal of performance issues during early growth. Not only was our user-base growing quickly, the nature of our software (payrolling and HR) meant that payroll processing periods at the end of the month would all line up, putting huge amounts of pressure on the software.

These two factors made it a continuous struggle to stay a step ahead, ready for the next peak moment of use. Unfortunately, our customers were often complaining that the application was slow before our internal teams noticed. Our infrastructure team was responsible for monitoring the application, it was their job to ensure we would see alerts if the application slowed, before users noticed anything severe. In many cases, users were still a step ahead.

Eventually, we noticed that our monitors and triggers were set up around very technical metrics such as server CPU, memory, disk IO, etc. In theory, these metrics could relate to a slow application. High CPU on a server would normally indicate overload and would result in reduced speed. However, the application can often handle requests, even with high CPU percentages. What we lacked was user-related metrics and triggers, such as page-load for relevant parts of the application.

We set up baseline metrics and created triggers based on page load of relevant pages. Then, we could very quickly see how growth and users were affecting actual user experience, making it easier for us to act before the situation became critical.

Strategy Tip: Test for quality based on user experience, not just technical triggers.

Conclusion

Product development is intrinsically different for SaaS companies. Rather than a single round of development before bringing your product to the market, developing a quality SaaS entails continuous development, constant updates, and frequent releases designed to continue to deliver value. When product development stops, so does value.

The 5 pillars of the Vision to Value Framework supports this process, giving you the tools to create structure, processes, and teams capable of delivering consistent, quality, and seamless updates.

Development operations

"DevOps is not a goal, but a never-ending process of continual improvement."

– Jez Humble

Development Operations or DevOps encompasses the idea that Development and IT Operations have to work together. Creating a strong link between IT Operations and Development is crucial if you are to respond quickly to problems, create feedback loops from IT Operations to Development, and handle escalations and problems in a fast and practical way.

In this chapter, I will discuss some options for structuring DevOps so that you can maintain quality, monitoring, and are able to meet internal and external expectations.

DevOps

Traditional development necessitates carefully separating development and operations, but Agile often requires bringing them together. Agile gives ownership of a module or product to a single team, which means they don't just develop it, they keep it running. Here, teams must take on every aspect of usage and operations, bringing development, testing, and operations together. DevOps is one approach that helps achieve this by combining Development and Operations.

DevOps can be hugely beneficial to SaaS organizations, because it shifts attention from developing a product to running it. In standard product production, companies dedicate most of their efforts towards developing a product and then ignore it once it's running. Scrum and DevOps mean that you focus on the running product, where you can validate if users perceive what you are delivering as value.

DevOps - DevOps brings Development and Operations together, either merging teams to create smaller squads or creating cross-functional teams with

Operations specialists as part of development teams. Here, engineers can take control of the entire lifecycle of a module or product, handling development, testing, deployment, operations, long-term improvement, and maintenance. In most cases, DevOps utilizes a technology stack and tooling to automate processes, speeding up processes such as deploying code or provisioning infrastructure, which would require bottlenecks and waiting with traditional split Development and Operations teams.

DevOps can be very tricky to implement. Developers often work on their own and tend to isolate themselves to prevent disruptions. Most are very much in a mindset of continuous learning and development. Engineers come from a different school of thought and often don't develop at all. Bringing these two teams together can be challenging because you have to bring work cultures together.

At the same time, it's becoming more and more crucial for tech companies to merge Development and Operations. Operations could previously build infrastructure without paying attention to what developers were doing but this is no longer the case. Automation is becoming a standard and may be essential for scaling. You may not be able to keep up with manual infrastructure technologies. Infrastructure and operations have to become part of development so that operations can build infrastructure for automation and for the application, otherwise it simply will not work. Mixing operations with development will enable you to quickly roll out new infrastructure to meet application needs, which will allow for faster scaling.

DevSecOps - DevSecOps is something of a "next step" for DevOps, in that it integrates Security and often Quality Assurance into Development teams. This can be extremely important for SaaS applications, where security must be considered, but where maintaining a separate security team would create bottlenecks and slowed development.

Here, your main goal is to push the idea that everyone is responsible for security. If your squads are responsible for an end-to-end module, they also have to be responsible for functional requirements such as security. Why? Security is extremely difficult to implement externally. At the same time, most security flaws are application related, so it makes sense for the developing squad to handle security as well.

DevSecOps allows squads to take on more independence and to scale more quickly. Like DevOps, it's difficult to implement because you will have to bring experts from different backgrounds and work cultures together. Doing so will require driving interest in operations and security for developers, development for operations, and so on.

Strategy Tip: Integrate development and operations to connect development teams to end-users to improve maintenance, operations, and deployment.

Service Level Agreements

SLAs are a necessary part of most customer-facing software companies because they define commitments to your customers. They let users know what to expect when something goes wrong, so they can make judgement calls based on how well software and quality standards meet their needs. This is important because many SaaS companies rely on strong relationships and trust, which necessitates communicating guidelines for solving problems, reacting to code red and code black issues, and guiding expectations for customer service.

At the same time, many organizations adopt a "set and forget" policy for SLAs. They're listed on a website and in some internal documentation, but not really integrated into business processes and actual work. This often means there is a gap between what external SLAs promise and what teams are doing – without teams even being aware of the gap.

The result is often that teams are disconnected and cannot track work to SLAs. Teams might fail to meet SLAs or breach them in one way or another because they don't understand what they should be doing. The answer is to create a structure for SLAs inside business processes, so that teams have to work according to the SLA.

If you want SLAs to remain a valuable and functioning part of your processes, they must be linked to triggers and monitoring throughout your organization.

Here, an SLA defines a standard which is further defined by KPIs or indicators. You must meet these standards to maintain a specific level of quality, customer satisfaction, the services you deliver, response times, and even how you're measuring that quality and response time. Linking this data to business processes helps you add internal and external value, because SLAs define how you are doing, how you are meeting customer expectations, and when you aren't. In this way, SLAs are most valuable when used as a boundary or trigger. When you no longer meet SLA standards, you must act.

Your internal SLAs are also important, but should be separate and tighter than those presented to your users. This will ensure that development teams have a buffer for fixing problems in case they don't meet expected deadlines.

Measuring internal and external SLAs means setting KPIs that track whether you are meeting guarantees listed in SLAs. For example, if you've defined uptime, call resolution, and time-to-recover after outages, you have to choose KPIs that tell you if you're meeting those standards.

Managing SLAs is difficult because it entails reviewing a great deal of raw data. To do so effectively, you must define what represents success across each of your SLAs, so that you can determine if they are being met or not.

Metrics such as customer response times, severity of incidents, physical damage to servers, etc. can all impact time investment, quality, and actual investment. This data will help you create a bigger picture of your customer service team's performance. Why? Different situations require different reactions and you won't have a clear picture of what's happening without all that data. Most reports aren't flexible enough to integrate why something may have taken longer because they look at specific types of data rather than the bigger picture.

Let's say two customers contact customer support with the same question. Customer A emails every 24 hours. Customer B emails every hour. Service Representative 1, who is working with Customer A, solves the support request in 3 days. Service Representative 2, who is working with Customer B, solves the request in 2 days. A KPI that only looks at time to close requests would state that Service Representative 2 was more efficient, despite them needing over 10 times the communication to solve the same problem.

Creating SLA tracking and monitoring covering all available metrics allows you to see what is impacting quality of service, what you can do about it, and how to keep improving rather than valuing the fastest or best regardless of circumstances.

You should also take some time to define how to set those standards, so that you can meet them. If you choose arbitrary times based on customer expectations, your SLAs may not align with business goals or even be feasible for a large part of the organization. Setting SLAs means defining a balance between customer expectations (As high quality/as fast as possible) and achievable metrics inside the organization.

Your first step should be setting a baseline with these metrics. This may mean defining SLA metrics against nothing but expectations and previous performance inside the organization. It may also mean reviewing how you're performing against current SLAs. Next, you would want to review how those SLAs align with organizational goals as well as what the customer sees as "value". It's also valuable to get input from customers. What are you doing well? What are they not satisfied with? What could you do better? While the obvious customer answer is usually "Faster, better, more "customer input can be extremely valuable in defining expectations and working to achieve service levels that add value for the customer.

SLAs must also be sensitive to the case at hand. Not all incidents should be treated equally. A customer asking for basic information doesn't require the same response as a customer reporting a major issue. A customer on a budget version of your app doesn't require the same level of support as an enterprise client.

A good approach is to define several incidents and then define their severity. For example, what is critical, what is low priority, and what is normal? This will help you to create a very clear matrix of examples. You can then use existing data to define how long something will take, how quickly you can respond, and what you can expect your teams to achieve at each level of support. For example, if you promise 99% uptime at all times, you should have historical data behind this. You should also know how long it takes you to get your network back up and running after an outage, and have the means in place to do so, even if your servers are completely out.

Having a support matrix of this type also allows customer support including first and second line to escalate problems into a relevant category based on actual severity of the problem without involving third line support.

Creating service level agreements of this type allows you to offer better support to customers, in that you can share defined periods in which they will receive an answer. It also allows you to improve the quality of customer support by ensuring that all tickets are given proper priority.

Finally, it's important to integrate monitors and triggers to track when something goes out of the SLA, so you can monitor quality.

Strategy Tip: Align external SLAs with internal processes and metrics.

Internal SLA

Internal Service Level Agreements are an important element of escalation because they define soft reaction times. Here, you must define what qualifies as an escalation, its level of security risk, and develop clear guidelines for an acceptable solve-time. The best practice is normally to establish different types of escalations – such as security risks, performance problems, or application outages – and then define solve-times based on the escalation. These solve times will likely be shared with customers and will be viewed as agreements.

If these Internal Service Level Agreements are not in place, teams may plan to do work, but may not prioritize solving problems as quickly as they should, because they won't have a clearly defined timeline. Setting expectations for escalation resolution will ensure that problems are fixed as quickly as possible.

Internal SLA's should be linked to external SLAs. While some organizations refuse to create external SLAs, this can backfire in that they often don't create SLAs at all, and tech teams may be unaware of how quickly they should react to critical events.

Your internal SLAs can also help you define internal quality, including baselines and metrics. For example, if a squad is developing a web application, you should have defined baselines for acceptable page load times, internal metrics, and resolve times. This gives your squads a quality goal, so that they aren't just building things without realizing that they might be damaging overall quality and service level. You can then use them as a baseline for performance expectations, defining minimum standards of quality in terms of code, speed in responding to tickets, fix or resolve times, number of bugs or patches, outage periods, and so on. If reality falls below these baselines, quality is low. If it is well above those baselines, quality is likely high.

Here, your baselines should be recorded in documentation and shared across all teams. My recommendation is to either share everything in a single platform or to develop an internal SLA chapter, so that everyone knows what their guidelines are and what they have to commit to. You should involve relevant stakeholders, developers, the Product Owner, leads, and people with external commitments when setting those baselines.

Earlier in this book, I discussed how processes should be implemented in tooling. I make the same recommendation for using baselines for SLA for quality.

Simply creating documentation makes it very difficult to validate compliance. Implementing the process into tooling means you can simply operate at that level. For example, if you have a security issue, the time-to-solve and quality baselines should be in the product development platform.

Integrating internal SLAs also means that customer service and IT can naturally follow SLA processes. If customer support can see expected time to respond inside their tooling, they can more easily follow SLAs rather than having to manage them independently.

Finally, internal SLAs can serve as a simple quality guarantee. By creating baselines for quality and standards, you ensure that teams know what they are monitoring for. They can check compliance before customers even notice problems. If a customer mentions your service is down or it's too slow and you aren't already aware, your triggers are not properly in place.

Implementing these SLAs will give teams a guideline with which to operate, so that you produce higher-quality code, solve problems before customers notice, and are able to properly prioritize fixes when something does go wrong.

> **Strategy Tip:** Describe and implement internal SLAs regardless of external ones and share with teams.

Automating triggers

Monitoring production environments is crucial to ensuring and maintaining quality. This typically involves a process of collecting data, defining baselines, and monitoring thresholds to trigger reactions or alerts when things go wrong.

Your product development pipeline likely already collects and measures a great deal of data. You also likely have baseline metrics defining items such as critical thresholds for page load, infrastructure resources, etc. Most development tooling will collect data, and you likely already have people in place to monitor that data. However, you also have to ensure that these people see this data at the right time.

Doing so means creating a series of automated triggers, designed to send alerts when things go wrong or thresholds are reached. Automated triggers are crucial for long-term quality, because no team, no matter how dedicated, will constantly pay attention to data and metrics. No one is staring at dashboards all day to ensure nothing is going wrong. Over time, teams may even grow complacent and forget to check monitoring tools entirely.

Here, you first must define who the right people are. In most cases, the answer is the product owner, but depending on the situation and its severity, it might include an entire team or the organization. Your "ownership" matrix should link ownership of problems based on relevance, impact, and required action (for example, it wouldn't make sense to alert someone who can't actually solve the problem).

This matrix needs a process in place to ensure that information stays relevant, triggers are updated to alert different people as roles change, and that someone is always responsible for every product or potential issue. Your ownership matrix directly links to the escalation matrix.

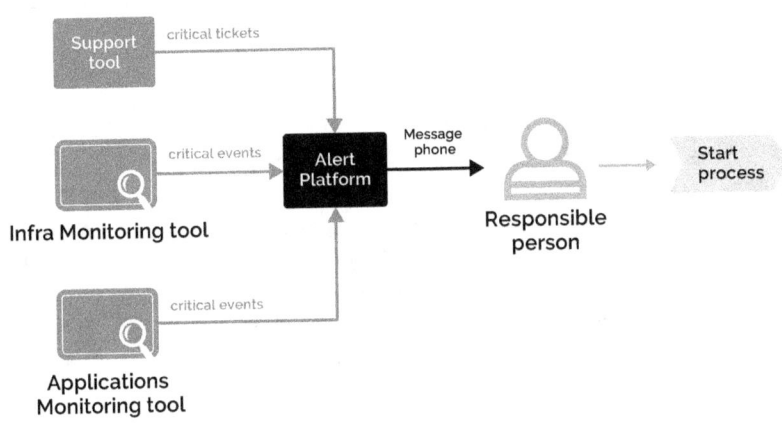

Fig.41. Triggers and alerting structure

It's also crucial to define how and when to share alerts. Sending too many alerts can be problematic because people eventually stop paying attention. Instead, you can define what constitutes as notification-worthy and send alerts for that critical instances. Everything else should be moved into a backlog for later review.

Creating processes where teams manually check dashboards once a day or once a week to review for problems should allow most small issues to be recognized and handled in an appropriate fashion. Here, teams can also look for issues and add new triggers as appropriate to continue optimizing the process. New triggers should also be added after code-red escalations, to prevent issues from happening again.

Finally, it's important to choose appropriate alert channels. You might consider using channels such as chat or Slack for small issues or minor alerts. You might also be able to automate adding items to a backlog for team review but shouldn't replace manual review. Major or critical issues should be linked to SMS or phone, because people will pay more attention.

Automated triggers ensure that people know when things go wrong without having to constantly pay attention to data. They also ensure you have a structured way to document when issues are noticed, which will aid documentation and compliance.

Strategy Tip: Only trigger alerts for critical events to avoid overwhelming product owners with notifications.

Your application never sleeps (24/7 Service)

Most organizations expect to operate 9-5 and then shut down. Everyone clocks out and goes home and no one checks back till the next day. Unfortunately, SaaS is not a 9-5 business. Your developers may work during business hours, but your customers and end-users are likely operating your software at all hours and in different time zones.

It's important that you define an operational structure mapping to when customers can access your product. For SaaS, that is 24/7. This means that you have to implement mechanisms to monitor quality and monitor for problems 24/7. These automated triggers should be set in place as early as possible so they function as part of your operations.

Offering 24/7 support for your application also means creating awareness and a culture of service inside the development team. If the shop is closed and everyone goes home, the application is still running. Your development team has to support that, keep people on call, and build triggers into the code to monitor everything so that disruptions and quality issues are noticed and fixed at any time.

This monitoring process should be connected to escalations, so that when things do go wrong, the right people are brought in. Here, you should have very clearly defined business processes, connecting individuals and roles to relevant cases, so that when something is triggered, the person in the monitoring role has a clear process and a defined person to call. If you have an escalation and don't know who to send it to, you cannot fix it immediately, which greatly limits the value of having monitors and triggers in place at all.

It's also very common for organizations to create triggers but fail to monitor them. If everyone just goes home, you have a control room that's been abandoned. You need processes to ensure someone is always monitoring triggers or that monitoring is automated.

At Nmbrs, we tackle 24/7 product support by keeping someone on standby, in a weekly rotating role. Their job is to pay attention, so that when something is triggered, such as a server is overloaded or goes down, that person can act on the trigger to solve it as quickly as possible. The person in the monitoring role handles a basic troubleshooting to determine if the case can wait until the next business day. If it's something urgent, they can escalate it to the right squad.

> **Strategy Tip:** Ensure your "castle" is always guarded, with 24/7 monitoring in place.

Escalations

Your development squads should naturally devote most of their time to development, with some room for operational work including checking and investigating quality, validating software, and performing routine maintenance. However, you should be able to disrupt that routine work when something urgent occurs so that the squad pauses all work to focus on a high-level problem. This is known as an "escalation".

You can think of an escalation like "fire". Everything stops, you have to fix the problem as quickly as possible, and you have to immediately know where to go. If you have a fire and can't find the fire extinguisher, the problem just gets worse.

It's important to develop strict business processes defining escalations, which teams or squads are responsible in each case, and how long squads have to solve the problem. Creating an escalation matrix means identifying internal products, connecting them to squads, and then developing triggers connected to those products. This ensures that when something is escalated, it automatically goes to the right squad. With an escalation matrix, you are essentially creating a fire drill, validating it, and ensuring it can run as quickly as possible.

It's important to keep in mind that things will always go wrong. Imagine a new employee becomes responsible for an escalation. If his phone number isn't yet in the matrix and you cannot reach him at night, you cannot solve the escalation. Or, your escalation matrix may not be properly set up and you might not know who's responsible. Taking the time to invest in a strong matrix that clearly identifies ownership, captures required information from responsible parties during onboarding (at-home contact details, etc.), and forces updating as part of its process is crucial to ensuring the long-term success of your matrix.

It's also important to define several levels of escalation. Here, the first level of escalation is a problem which can be escalated within a certain squad or team. First level escalations should be solvable by a single squad because they are in the scope of that squad and that squad should have everything to solve the problem. At Nmbrs, we defined this escalation level as code red. A Code Red overrides all other planned work being performed by the squad so they can solve the problem as quickly as possible.

Second Level escalations are more serious and likely impact the business. This might mean the application is down, performance is too slow, or a critical feature is not working. Here, you need to escalate the problem to the company, including account managers, customer service, and likely stakeholders as well. The same teams (who are responsible for the parts of the application not working) will still complete the work, but the rest of the organization can operate around the problem. For example, management can work to manage the business risk, customer service can communicate the problem to customers, and account managers can communicate to individual account holders.

Escalations create clear processes on how to handle urgent situations, such as security or data leaks, application downtime, or core-functionality breakage. In these cases, you should include communication as part of the process, because communication is one of the most important things you can do. Ensure the fix team is communicating internally at each step, even if there is no clear solution, so that support teams can communicate externally to users and relevant stakeholders.

Finally, it's important to follow up on escalations. Once you've put the "fire" out and have the situation under control, you still want to learn from it. Create processes to ensure retrospectives are organized and the people involved can discuss root causes, solutions, and what could prevent the same or similar incidents in the future. These retrospective sessions should be organized by a facilitator such as a scrum master to ensure the right questions are asked and that the core problem is found. The most important part of the process is to learn from the crisis so that you can improve your product and process quality.

Strategy Tip: Define different levels of escalations with clearly defined owners.

Conclusion

Having DevOps in place supports scaling, because it enables teams to own end-to-end processes without bottlenecks or coordination issues. Laying the groundwork for strong support, creating processes to manage and handle SLAs, and implementing processes to manage quality and value will help you as you begin to scale and need these processes.

At the same time, defining internal and external expectations in terms of service level agreements, customer expectations, and how you handle and manage escalations and code red issues will help you at every point. Taking the time to define these things will tell you where you want to be, why you want to be there, and how to get there, so that you can build structure and teams around achieving desired quality and customer value.

Customer Support

"Your most unhappy customers are your greatest source of learning."

– Bill Gates

Customer support functions to ensure end-users see value from your application. Good customer support is the most basic requirement for providing quality service to your users. As COO, it is your responsibility to structure support teams, but how you do so will impact how customer support is able to function, support knowledge of the application, and how customer support is able to directly impact feature development.

In most cases, customer support is your most direct line of contact with end-users. Customer support teams connect to end-users, knows what they want, the problems they experience using the product, and where the application is and is not delivering value. Many SaaS products integrate customer support into the app, so support is very much part of the product.

Customer support impacts development intertwines with the product as part of value, and – for a SaaS company which sells a service as much as a product – is essential to delivering value to consumers. Customer support must be linked with development, so that the team has a very strong working knowledge of the application, its flaws and bugs, and how it works.

Most companies will want to handle customer support in their own way, which is good. However, in this section, I will go over best practices for structuring and realizing customer support to sustainably scale operations while maintaining quality of customer experience.

Define your support identity

Customer support is your direct line to customers. Service agents represent the organization to consumers, becoming its voice, acting out its policies, and creating customer opinion of the brand. Defining your support identity is crucial to defining your organization's brand.

There are many ways to approach customer support, ranging from self-service to full service, with communication ranging from casual to formal. Taking the time to define your customer support identity will help you offer better service.

Customer support has traditionally been viewed as unnecessary and problematic. This naturally led to many top organizations developing a reputation for poor support, difficult-to-find contact information, and even unhelpful support staff. Companies like eBay, PayPal, Comcast, and AT&T maintain that reputation to this day.

Today, customer expectations are changing. Individuals expect and demand friendly, helpful, fast, and personalized support through a range of channels such as social media, phone, and a direct line of chat or ticketing in your application. Failing to offer quality customer-facing support hurts your ability to offer value to customers, and therefore your ability to retain customers.

Defining a customer support identity is one of the first and easiest ways to offer that support.

SaaS companies typically provide both self-service and full-service, with a focus on one of these styles. Here, self-service encompasses customer support where the end-user finds and actualizes as many of their own solutions as possible. Customers utilize forums, knowledge bases, FAQs, and tutorials to solve most of their own problems. Here, end-users only contact customer support for help when they cannot solve problems through self-service resources.

Full-service customer service is much more hands on, where you directly interact with consumers and actively work to solve their problems. Customers contact a helpdesk, chat, social media channel, phone line, email, or messaging to receive direct and hands-on help.

This type of customer support often means permanently assigning customer support agents to tickets so that individuals stay in contact with the same person as much as possible. It means actively communicating back and forth with the customer while solving problems, and it means staying with the customer until they've resolved the problem.

There are pros and cons to each, and you will have to consider several factors such as response style, your demographic, and the quality of end results when making this decision.

For example, your customer support response style will determine how and when customer support talk to customers. In some types of full-service customer support, teams write out long and complex tutorial-like answers. More modern forms of customer support often revolve around creating a dialogue with consumers and remaining online while they work to fix the problem. Not all customers will appreciate a chat-based approach, not everyone will appreciate long tutorials. You have to structure your support identity around your customers.

You also have to consider the voice and tone customer service are expected to use. Are they speaking formally with your end-users? Are you creating a more informal and friendly brand image? Your communication guidelines should be in place to guide the formality of conversation, but shouldn't be so prescriptive that customer-service agents respond in the same way every time. Your tone of voice and language usage should be chosen to match company identity and branding, so customers experience a seamless identity across the organization.

Another important factor is the quality of support. While you obviously want to define a customer support strategy offering the highest quality possible, there is a point where quality ceases to add value. Here, you have to consider factors such as speed of customer support (how quickly do you reply? How quickly do you reply in each time zone?), customer support team structure (how many resources do customer support teams have for factors such as onboarding or troubleshooting), and processes (how do you define problems and emergencies? Do you give customer support autonomy to actively fix problems on their own or escalate them to someone who can fix the problem? Have you created processes to communicate problems to developers, escalate issues, or even refund customers?).

If you focus on offering the best quality answer as possible but take too long to answer it, the customer might not receive any value from that answer. Or, if you offer strong customer support for existing users but have no resources to offer quality onboarding, customer-service resources are being poorly spent. Your best option is to balance your resources to provide the most value in every aspect including onboarding, troubleshooting, problems, and emergencies, rather than focusing on one.

Each of these factors will play a large role in your customer experience. It's important to define them when setting up your customer support. For example, your business processes relating to customer support will directly define how customers are helped, the autonomy customer support receives in solving issues, and how they are able to directly share customer information and support with development and other teams.

Strategy Tip: Define your customer service identity and structure support around it to offer a consistent experience to end-users.

Customer support structure

Your customer support structure defines how and where end-users are able to check for and receive information they need. It also defines the priority at-which you handle incoming requests by creating levels of support, defines the resources used to offer support, and defines a structure connecting what's happening inside the organization to customer support. This structure will involve creating technical resources and documentation as well as strategy and business processes to guide customer support.

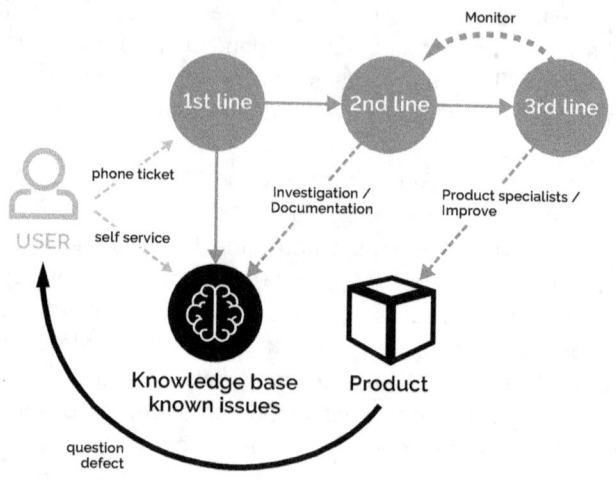

Fig.42. Customer support structure with 3 layers

Developing support lines is one of the first and most important steps you can take. Most technical companies create three lines of support. Here, first line filters basic questions which can be answered with a knowledge-base or general knowledge of the application. These questions aren't necessarily indicative of a problem, only that the user is not yet familiar with the software and is making user-errors. First-line support can also handle queries relating to known bugs and errors, which have already been escalated and are in the process of being fixed.

Second-line support investigates problems and troubleshoots, typically serving as a filter to ensure that problems identified by first-line support are actually problems. If customers encounter an unknown bug, if something breaks, or if you experience another unknown problem, second line can step in to investigate the problem to determine if it's a known issue and how to fix it. If they cannot fix it, they escalate to third line.

Third line support are technical experts who are either part of development teams or who work closely with them. Third line is technical support, who should be capable of solving technical problems including bugs.

This multi-tiered system works to reduce the total cost of customer support by reducing the number of technical experts needed on the team. First and second line exists to keep third line from spending time fulfilling basic requests, while real problems go unanswered. Developing this type of system also allows you to offer customer support on an as-needed basis. Here, some customers will largely fix their own issues and won't need support until they're ready to move to second or third line. Other customers will want support with every issue, even when it's available in the knowledge base and can contact first line.

However, this customer support structure becomes most interesting and useful when connected directly to development. Why? Third line will ideally function as a communication channel, so that customer support always knows what development is doing and development always knows what customers are saying.

How does that work in practice? Organizations are more and more often integrating customer support directly into development teams to create a direct communication channel with customers. Doing so allows developers to directly see what customers are asking and why, with the context delivered by customer support. It also enables customer support to stay on top of changes, so they can offer the best possible support.

For example, Atlassian[23], an enterprise software company, experienced problems relating to a lack of visibility for bugs after problems were raised, zero-context for developers regarding bugs and customer complaints, and the lack of a feedback loop for both teams. Atlassian attempted to create a custom team to solve the problem. They failed to do so until they integrated Development and Customer Support onto the same platform, enabling both to work together. Their new system ties development into directly to customer support, so that support representatives can see the "fix" status on existing tickets, can search old tickets based on problems, and can directly escalate tickets. It also immediately shares updates – for example, when a bug has been fixed or a patch has been released – to customer support

Buffer, a social media management platform, takes this idea a step further by integrating customer support into development teams. Buffer's average team structure[24] comprises 1-2 engineers, a product specialist, a customer development representative, and (normally) a growth analyst.

This growing trend of creating direct lines of communication between development and support is an important one because it allows you to make several key decisions for your customer support.

The first is that third line support will always know what's going on in development. Third line is aware of development schedules, releases, rollouts, and recognized bugs and issues. They already know when the product will change, when new features will launch, and when the knowledge base has to change. Third line can then automatically update the knowledge-base and inform first and second line of upcoming changes. Customers will often seek answers in the knowledge-base and first line works from the knowledge base, so keeping it up to date will improve the quality of customer support while reducing unnecessary-escalations to second line.

This also positively benefits customer support in that first line will already know why they're getting new tickets. If they are aware of changes, they can easily pass that knowledge on to customers with limited frustration on either side.

Finally, third line has to work with developers to find and solve root-cause

23 https://www.atlassian.com/blog/software-teams/tearing-walls-development-support
24 https://open.buffer.com/product-team-evolution/

issues. Solving the underlying problems behind support tickets is key to driving value from customer support, so that you resolve issues instead of temporarily solving customer problems.

Why is this important? One of your biggest risks when scaling is that second line will simply troubleshoot and solve tickets without connecting them and without solving the root issue. Instead, they'll keep solving tickets individually and will keep getting more and more questions. Customers will keep sending more tickets because they'll continue to run into the same issues.

For this reason, I would strongly recommend skipping second-line support for as long as possible, so that third line can immediately solve root issues. You will eventually need second-line support to filter non-priority tickets, to group different types of tickets together, and to prevent third line from becoming overwhelmed, but you likely won't need second line until you are well on your way to scaling. When do you introduce second line? If third line cannot keep up, second line works as a good way to filter customers and to collect more information for third line, so that they can focus on actually solving issues rather than on collecting information and troubleshooting.

It's also important to establish a structure allowing questions to be grouped together. Second line won't always recognize when problems are the same because questions are different and they are not technical experts. Creating a database so that questions can be grouped and the root problem or cause can be identified may be difficult, but it's a good idea to introduce a system to do so.

Your customer support structure should also define how you approach security and customer privacy. For example, if someone calls and asks for a password reset for an email, how do you verify they are the account holder?

First-line support needs to have the resources and ability to identify who is communicating to them. This can be established using business processes inside tooling, so that verification processes are automatically followed as part of connecting with the customer.

Every interaction should be registered and mapped to a ticket, so you can track customer contact per ticket, number of tickets per user, and how different segments or demographics use customer support. This will allow you to improve customer support by mapping support tickets to larger root-cause problems, to analyze customer support needs per demographic, to analyze the quality of support, and to work to reduce the total number of tickets per customer.

Customer support is extremely important, but no one contacts support because they want to. One of the largest goals of customer support should be to recognize root-cause issues and to raise those concerns with development, so that customers have fewer questions and tickets, therefore raising the customer's perception of application quality.

Strategy Tip: Connect customer support to development to create a two-way communication channel.

Support Channels

Creating customer support means choosing how and when you want to offer support. You must create a balance between self-service information (e.g. knowledge-base) and direct support (e.g. phone, chat, email, or helpdesk).

You will have many different types of customers, who each want and need different types of support. Some customers will prefer to call and others will prefer chat so you need more than one type of channel.

You'll also have to determine the purpose, the level of support available, and how to define expectations on each channel. It's important not to have too many channels because they will become difficult to manage. Your best option is to review which types of communication are relevant to the application, service, and support line. For example, a chatbot will not offer quality support for complex questions relating to your application. On the other hand, if you're filtering basic questions, a chatbot could be very useful.

Your goal should be to identify which channels make the most sense for your application. Once you have channels selected, you need structure for organization. Questions in any channel should be registered as a ticket, so that customers can escalate problems inside of it, and so that quality and security can be monitored.

Channel	Pros	Cons	Best For:
Ticket via Email	Easy for customers to send Good for open questions	Information is unstructured and customers will rarely send everything you need at once, so support agents will have to parse it. Users may have to wait for an answer, which may be too long for simple queries	Open questions, longer or more complex problems
Ticket form	Data is structured. You can ask the customer to submit relevant data fields, such as actions performed, error codes, etc. Good for simple and standard questions or issues.	Support may feel less personal Users may have to wait for an answer, which may be too long for simple queries	Longer or more complex problems, Troubleshooting
Phone	Easy for customers to quickly get support and answers Users can have simple questions and problems solved on the spot	The agent on call may not be able to solve the problem and may have to escalate to another support agent. This can result in delays, long calls, or an unsuccessful support call	Quick solutions for simple questions or known problems
Chat	Gives the user a chance to ask open questions. The customer receives real-time support.	It may take many interactions to reach a solution, especially with more complex problems	Quick solutions for simple questions and known problems

Taking the time to define channels and how they work can also help you to define a structure and processes for support. For example, if you know that third line support cannot solve issues over the phone, you can ensure that customers or first/second-line agents have to submit a support ticket to escalate rather than transferring the problem over a call.

Strategy Tip: Only use support channels that are relevant for your users.

Knowledge structure

Most people don't want to contact customer support. While there will always be exceptions, your average customer will likely prefer to solve problems as quickly as possible and get back to using the application. For most, this means solving the problem themselves. Most users will only contact customer support when they're actually stuck. This is actually beneficial for you. It's much more cost-effective to have users refer to a knowledge base and to solve their own problems than to continue contacting you for simple and common issues.

Creating a strong knowledge base will also benefit your organization in other ways. For example, your first-line support can use it as a reference when helping customers. Your knowledge structure or base should provide information on basic use, application interface, and common problems or bugs, even those that are being patched or fixed.

Here, it's important to define what types of information you would like to include, how in-depth that information should be, and what audience that information is directed towards. It may also be beneficial to set up several knowledge base directories, so that information is organized at different levels of technical capability.

Your knowledge center should be structured around product knowledge, with information on setup, use, troubleshooting, and learning. This can incorporate more in-depth components such as training, e-learning, complete walkthroughs, and more. It is something of the reference book or manual for your application and you can choose to add as much or as little information as necessary. It's also important to consider the user, their goals, their stage in your application, and their technical knowledge or capability when creating your knowledge base.

Type of Resource	When to use
Article	Detailed explanation of complex features work, reasoning behind it, how calculations might work, etc.
FAQ	Quick answers to frequent questions. Longer explanations may link to articles.
Troubleshoot	Symptom/actions list where users should be able to match the problem they're having and where possible solutions are offered. This is a good asset for your 1st line support because they can help customers quickly.
E-Learning	Explain the overall concept behind features. E-learning material can also create learning paths so that users can track progress.
Webinar	When you want to cover complex topics and need interaction from participants, but still you want to cover a broad audience.
Walkthrough	A good option for in-app guidance, including how-to information, feature setup, and product tours
Live training	Train people for complex features and topics. Offers a close connection with participants, and can be adjusted on-demand but is normally sent to a limited audience

Most SaaS organizations greatly benefit from integrating information explaining how the product works, how to set up the product, and basic account and interface actions. This information is extremely valuable to beginning users and should often be available with picture or video tutorials guiding new users through the process. Having this information becomes more important as the complexity of your application increases, but once your application exceeds a certain complexity, it becomes irrelevant.

For example, if your onboarding process requires significant technical expertise and setup, you would want to assign an account manager to help with onboarding process anyway, so offering knowledge base information on how to do so would be redundant.

You'll also likely want to include a troubleshooting section, where you define common and recognized problems. Here, interested users can simply search their problems, track it to a common issue, and see either a solution, the status on the solution, or another fix.

Known-issues sections offer value when helping users solve problems and when updating users to the status of unsolved problems. If you are in the process of fixing a bug, adding it to a known-issues section with a note that it has been recognized and is being fixed will save you a lot of tickets. This will require a close collaboration between third line support and development, but if you've structured your support teams to include that, you'll already have that communication channel open.

Finally, your knowledge base functions in the same way as software product. It's only relevant and useful if people are looking it up and continuing to use it. If data is out of date, not updated, or not searched, it is not adding value. Your knowledge base should have clearly defined owners (typically per section), so that all relevant squads can contribute to it in an orderly way. You may want to define a structure allowing support squads to directly update the knowledge-base themselves (not highly recommended for customer-facing content) or define a structure where squads pass information directly to third line support, which then goes on to update the knowledge-base.

Having this structure will ensure information is kept up to date, that it is always shared in the same way, and that information always goes to the right place.

> **Strategy Tip:** Define an audience and purpose for your knowledge base and structure it accordingly.

In-app support

Most knowledge base information, including setup and information about how the product works, should be integrated into the product itself. To achieve this, customer support must collaborate with developers to integrate setup and tutorial information in the application.

Why? If an end-user has to visit a forum or a knowledge center, something has already gone wrong. If they aren't getting answers for standard questions during normal product usage, there's already a product defect.

Establishing integrated knowledge base information throughout your application requires development to establish a close relationship with customer support. Customer support creates a feedback loop, establishing where and how customers use the product, where they experience difficulties, and what questions they have, so that development can update the application to simplify, to include explanations, or to launch tutorials for processes that must remain complex.

Linking that knowledge base directly into a product will save customers time and make information significantly more accessible. Why? As a user, the very large amount of information in a knowledge base can be frustrating. They have to know what to search for, how to search for it, and where or how to access it. Some applications only make information available using specific terms, which the user may not be familiar with.

So, if the user has to search for information through a knowledge base, it is disrupting application use. And, if customers have to do so to figure out how the product works, the product has poor usability or high complexity in some features.

Consider a product like an iPhone. Users don't often have to look up how to perform basic actions such as calling a contact or sending a message because everything is intuitive or explained. Usability drops in the case of more complex features, where users may have to search the Internet for answers.

Closing this gap and reducing the need for users to look up information means integrating in-app support. Here, users are shown how the product works the first time they access the feature or are directly linked to the knowledge base without the extra step of searching.

When Nmbrs integrated in-app support, it greatly reduced the number of incoming tickets and phone calls, and increased user satisfaction.

Strategy Tip: Structure support information so that it is delivered alongside product features as end-users need it.

Product Communication

Your customer support team is your only direct line of communication with customers and end users. This means that your customer support team is the most logical and efficient way to communicate product releases and updates to consumers, because they have the open channels, defined ways to connect with customers, and established relationships to do so. In addition, pushing product release information to customer support first means that customer support will have the information they need to answer incoming queries and questions related to the update or new release, can help to track customer reception, and can help to monitor quality.

It's important to define a process to push product release information to customer support. Customer support should also have a defined process for moving that information to the end-user.

Here, your product communication will change depending on the product and product type. Release notes are the easiest way to communicate information for new products and features. Your process should define how customer support receives release notes, how they use this information to update databases and knowledge bases, and what channels they should share information on.

Your communication channels should share both negative and positive information. Most organizations are happy to share new information relating to product and feature releases, but significantly more reluctant to do so when it involves problems. However, sharing this information and keeping users informed will reduce total demand on customer support, will reduce duplicate tickets, and will boost customer satisfaction. Creating a database of known issues and their known workarounds or fixes will prevent customers from having to contact customer support, only to be told you're aware of the issue and cannot fix it now.

It's also a good idea to set up a status platform, so that if part or all your application is down, you can easily communicate it. Sharing the platform status and progress towards fixing issues will reduce the demand on customer support, so users don't have to call to learn about update or repair status for their product.

I also recommend developing this sort of product communication as part of the application, so users can see status, updates, and known issues during product use. This ensures users don't have to lookup information because it will be pushed wherever relevant.

Strategy Tip: Integrate status updates and product updates into your application.

Scalable Support

Scaling your organization means increasing the side of your customer base, which naturally results in those customers having more requests and more questions for customer service. This can be problematic for customer service teams who don't have processes in place to handle increases in ticket volume.

Customer support will often approach tickets on a one-by-one basis, working to solve each as quickly as possible so they can move on to the next. This can be a mistake. It forces customer support to see tickets individually rather than as interrelated, even though the latter is often the case. Most tickets fall into groups that revolve around the same issues.

Changing this requires developing processes and an internal mindset where customer-support agents work to analyze tickets patterns and their root-cause, rather than closing individual tickets. Even first line should be asking questions like, "Why is the customer asking the question?" If you can identify why tickets are coming in, you can group them, find answers more quickly (because chances are the ticket has already been solved) and escalate the root issue to development, so they can work to fix it.

When customer support answers questions rather than identifying root causes, they will keep putting the same effort into answering problems they've already solved.

If you can't identify when questions refer to the same problems or issues, you'll need a very large team to handle every incoming question. If customer support agents feel like they frequently receive unique questions, solving those problems will take a long time. You cannot scale, because adding more unique questions from more customers will overload your teams and you won't be able to keep up without increasing the size of teams. If you approach every question as unique, the more tickets you solve, the more you will receive, because the same problems will continue to arise.

No ticket is unique. There will be a certain number of possible questions related to your product (although this can exceed thousands or even tens of thousands) with some logically being more common than others.

The best practice here is to create a database or framework that clusters questions around their root-cause or problem. Once you do so, you'll quickly see that most questions revolve around the same problems.

Organizing questions in this way means you can look at inbound questions in terms of what's already been solved. If a question tracks to an existing root-cause, customer service can very quickly solve the problem. If not, they can escalate immediately.

This type of framework creates a repeatable and scalable model for customer support. When no ticket is unique, solving most tickets will be a process of searching the database, finding the solution, and sharing it. This means you can serve more customers with fewer customer support agents.

While this will require more time and investment upfront, it will save money over time.

Using a framework to recognize root-cause issues can enable flows to minimize the number of tickets per customer. If you can cluster tickets around specific issues, you can measure the quality of specific features and aspects of your app. This tells you where to improve, what to change, and what customers are concerned about, so developers can actively tackle problems in future updates.

You can also use root-cause frameworks to identify the most common issues in your application and therefore to update your knowledge-base, so customers can use self-service. At the same time, relying too much on self-service will reduce customer intimacy, so you should never rely solely on databases, even for very common problems.

> **Strategy Tip:** Create a mindset where customer service looks for the root-cause of incoming requests.

Bridging the Knowledge Gap

Learning is a natural part of trying something new. Anytime a new customer begins to use your product or application, they will do so with no or very limited foreknowledge of the product. Bridging those knowledge gaps can be simple, such as when software is very familiar and based around commonly used interfaces. It may also be extremely complex with a steep learning curve. In either case, your customers and users will likely need a certain level of support and training throughout, so they can quickly adapt to your application and begin realizing value from it.

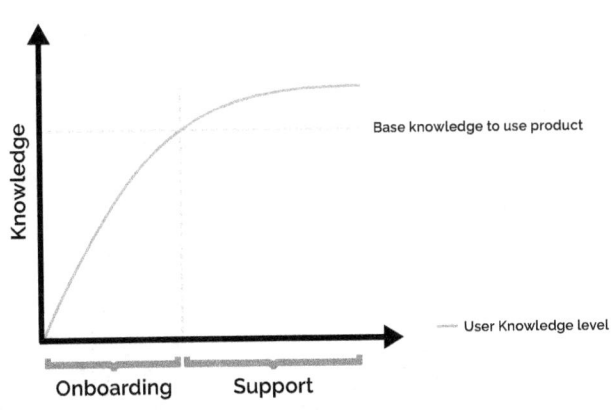

Fig.43. Growing the user's knowledge in your product

Your customer support team is responsible for analyzing and recognizing knowledge gaps, so they can provide the necessary training. Why? If users don't have basic training or knowledge, they'll either become frustrated and quit or will stay in customer support asking basic questions. My recommendation is to develop training material and integrate it into the app. Doing so will improve the average customer experience. It also ensures customer service can continue to function as a maintenance team, rather than spending all their resources on onboarding.

Developing onboarding training means understanding how individuals approach your product, their technical level, and what they need. In the case of very complex products, you might need a full e-learning program, on location training, or webinars. You may also need simple tutorials worked into the application, so that users can find their way around your product more quickly.

In most cases, you'll also have to account for users at different levels. Some users will come from similar programs, will be very technically savvy, or will otherwise pick your program up quickly. Others will require much more in the way of assistance and support, which will mean more investment on your part.

You have to ask questions such as:
- What steps are users taking when they launch this product for the first time?
- What programs are users coming from? How much of this app is new?
- How much of this app is intuitive to the average user?
- What do new users most often want to achieve first?
- What are the basic steps to complete to achieve full usability? How complex are those steps?
- Where can users get lost or confused?
- Where do existing users frequently voice frustration or ask for help?
- Which steps add complexity?

Answering these questions will give you a baseline for where and how to integrate information into the application, because they will help you identify where your knowledge gaps have the most impact.

It's also important to build training for different personas because you will always have different types of users at different levels.

Here, e-learning modules are the most scalable solution, because you can serve every customer with no added cost. Developing e-learning at a level appropriate for most customers is usually a smart investment. However, you will lose the ability to offer a tailored experience and ask in-depth questions about what the user doesn't know or what they need the product for.

If your product is very complex, requires significant personalization, or is otherwise very technical, you also have to include customer support information or product how-to information in the product. It may be better to offer webinars and in-person training to ensure users have the required knowledge to gain value from your product

Why would you want to integrate customer support into the product? The more difficult your product or information handled by your product, the larger the knowledge gap you can expect. You are aware of what that knowledge gap is or should be for most customers, and easily preempt questions and frustration.

For example, QuickBooks is a relatively complex SaaS. New users have to take numerous steps to create their account and get everything working. If they don't, they can't fully utilize the software. So, QuickBooks integrates a "User onboard" feature that walks new users through processes like connecting a bank account, invoicing, importing data and files, paying employees, and so on[25]. These tutorials are built into the app, show up on the home page, and allow the user to click on them at their own pace and in their own time. QuickBooks also offers more in-depth onboarding for customers who need more support, including the option to have dedicated account managers. eLearning, regular webinars, and other types of learning are also available for specific situations.

Investing in a great deal of customer support and information, most of which is integrated into the app, saves QuickBooks a lot in terms of customer frustration and by ensuring customers can make the most of their product.

Other SaaS solutions like Canva offer simpler services and interfaces. Canva integrates a very basic "Tip" onboarding solution into the app, so that users can see basic walkthroughs when using services for the first time. Otherwise, they have to go to the knowledge base or search for answers online.

In most cases, a combination of models is the best solution. You can offer e-learning modules as a basic training resource, which anyone can access at any time. You can then add on webinars or in-person training and courses for customers in need of further support. Most importantly, this information should be integrated into the application, so that it's available where users need it.

If you do choose to use e-learning, it's important to create a channel so that users can ask in-depth questions, and that those questions can be forwarded to experts in third line support.

Nmbrs is a complex SaaS product aimed at expert users, primarily payroll administrators, who use it as a core tool for daily work. When these users change applications, they understand payroll concepts, but have to learn how those concepts are implemented in the new application.

While we strive to make the user experience as simple as possible, there is often a significant knowledge gap for new users. It takes some time before a new user is comfortable running our product.

Originally, we created a knowledge base detailing how the product works. We quickly noticed that most of our busy users didn't have the time to read long articles and detailed explanations. Knowledge base articles did not reduce the number of questions in our contact center.

We split our self-help, moving simple and basic use cases to an e-learning format. Longer and more complex questions, which were only needed at specific moments, remained in the Knowledge Base. We then integrated e-learning content into the onboarding process, so new users were automatically introduced when starting the project. This new process significantly improved customer

---

25 https://www.useronboard.com/how-quickbooks-onboards-new-users

experience, as many users preferred following e-learning to asking questions or contacting the support desk. The result was higher customer satisfaction, more engaged users, and a reduced volume of demand on support.

> **Strategy Tip:** Develop training material in the application so that you can quickly help new customers bridge knowledge gaps.

Onboarding

Some applications require very complex or involved onboarding and will necessitate live customer support. For example, if your application requires that customers import a great deal of data or accounts, offering live assistance in the form of an onboarding team will help things to go more smoothly. It also ensures even non-technical users have correctly set up their account so that they can see value from your application.

Developing an onboarding team means creating an element of customer support solely for the purpose of introducing customers to the platform, for solving initial customer questions, and to help customers with complex setup processes. It's important that this team be closely connected to customer support and sales, so they can see where customers are coming from (other apps, knowledge level, etc.), where knowledge gaps exist, what they have to implement to meet the customer's needs, and how much support that particular customer is likely to need. You may choose to include your onboarding team in the scope of customer support, in the scope of sales, or as its own team.

While onboarding teams are a part of customer support, it's important to differentiate them because customer support is responsible for maintaining existing customers, while onboarding is only responsible for helping new users gain value from the app.

An onboarding team is not necessary for every SaaS. It is important if you own large customer contracts, have a complex onboarding process, or one in which customers frequently fail to properly set up their account.

Business and data-related apps almost always need an onboarding team because migration from a previous application is normally required. If customers fail to migrate properly, they won't realize value from your application. Investing in an onboarding team will boost customer satisfaction and customer retention because it ensures end-users get as much as possible from the app.

Your onboarding teams can also create a feedback loop to development, so that development can work to streamline complex areas and make future onboarding easier. The more easily individual users can onboard, the more scalable your product becomes. Why? If you need to have consultants handling onboarding, you're assuming your product is difficult to start. If you can streamline problem-areas and create a more DIY approach, your onboarding team can function in a guiding role, allowing them to onboard significantly more customers at once.

Onboarding brings sales and support services together, because customers are usually in the final stages of the purchase pipeline when you begin to onboard them. Your onboarding team should have access to resources from each, so develop your team accordingly.

At Nmbrs, we tackled this problem by developing an onboarding app, which offers training, documentation, and sales assets like demos and whitepapers. Our platform is HR and involves finance, so it's naturally very complex and customers often need support. Our onboarding team functions in a support role, collecting data and guiding new customers, but offering one-on-one support when needed.

Strategy Tip: Create an onboarding team aligned with sales and customer support to help customers achieve more value with your product.

Support Quality

Customer support exists to ensure that customers are receiving value from the product. Business processes and standards ensure customer support has the tools to achieve this.

The most important of these processes should be continuously improving the quality of support. Achieving this means defining metrics to measure quality. Here, defining customer satisfaction as a metric is not likely indicative of the success of customer support, simply because customers go to support when they have complaints and problems. Instead, you should define metrics customer support can control, such as the number of tickets per customer, the ratio of tickets per customer, and so on to track how well customer service solves root issues.

This typically means defining metrics pointing to quality of both service and content. Most organizations eventually need KPIs reflecting occupancy, customer satisfaction, and productivity.

Occupancy – Occupancy defines work hours spent on the core focus of a job, or what you want customer service representatives to be doing. This includes solving tickets, chatting with customers, answering the phone, and special products. It doesn't include maintenance work like employee training. In most cases, you want occupancy to be at about 75-80% for all core roles but you may want a higher rate for specific aspects of the role.

Productivity – Productivity helps to define what support representatives are doing with their work hours. It involves monitoring representatives, what they are doing, and how they are doing. Here, you want to define looked-for metrics including tickets solved per hour, tickets responded to per hour, and comments or chats per hour.

Quality – Assessing the quality of customer service can be difficult, because you can't always just ask customers. People contact customer support when they're upset, so they're likely to leave a bad review or share a negative experience. This can lead to misleading data showing bad customer service. Instead, you want to look at factors such as tickets per customer segment to indicate support quality.

You also want to create processes defining communication standards, so that you can assess if tickets are handled well. Consider factors such as communication, process, suggested solutions, and so on. Integrating this does mean implementing a peer review system where peers receive random tickets and discuss the quality. Using a peer review system, inspired in code peer reviews, prevents the issue of quality from being relegated to a single person. Instead, you want multiple people to offer insight and ideas into how the customer's experience could have improved.

Finally, you want to track how well the ticket followed guidelines, if it met the SLA, and if a root-cause solution was reached.

> **Strategy Tip:** Define quality expectations for customer service and implement a peer review system.

Conclusion

Customer support teams are your connection to the customer. They are the first to find out when customers aren't happy. They're often the first to know when something goes wrong. And, they have the power to communicate what customers are asking for, struggling with, and unhappy about.

Structuring customer support teams in ways that align with organizational goals is important. Customer service must be able to deliver support in ways that make sense, offer the most value, and that are scalable. At the same time, support teams should be able to create a feedback loop to development to create a process of continuous improvement, a backlog of desired features and updates, and a sustainable knowledge base which will benefit both development and support.

Governance
and Compliance

"You get a reputation for stability if you are stable for years."
– Mark Zuckerberg

Designing governance and compliance controls is one of the final steps in developing an organizational structure. Once everything is in place, you still want to be able to scale with as few problems or issues as possible. To achieve that, you have to put controls in place to ensure things continue to run smoothly.

You don't want to encounter major business problems because issues develop from a lack of compliance. Proper governance will work to keep foundations solid. As Tech COO, it is your responsibility to put mechanisms in place to guide the development process and ensure quality. IT governance should cover access management, change management, incident management, data security and periodical audits to validate that the organization and quality output is on the right track, merging IT with business goals.

IT Governance

As a small organization or startup, you will rarely have IT governance in place. As the organization grows, stakes will get higher and the risks of improper data access will grow. This can result in liability for your company in the form of data breaches for clients, lost data, or even corporate espionage. Implementing governance to manage these risks is crucial to scaling while protecting your customers, users, and goals.

There are many IT governance frameworks such as COBIT, ITIL, CMMI, and FAIR. Most focus on quantifying risk such as cybersecurity risk and operational risk and taking steps to reduce it. Most also help you to identify how your IT department is functioning, which key metrics you have to collect, and what you are getting in return.

No matter which framework you choose, your IT governance should include risk assessment and risk management. Here, risk assessment means specifically identifying risks in both long-term and short-term per business activity. Once these mechanisms are in place, they will benefit you in that it will be an easy step up to offer external quality assurance or to implement market standard certifications such as ISAE and ISO.

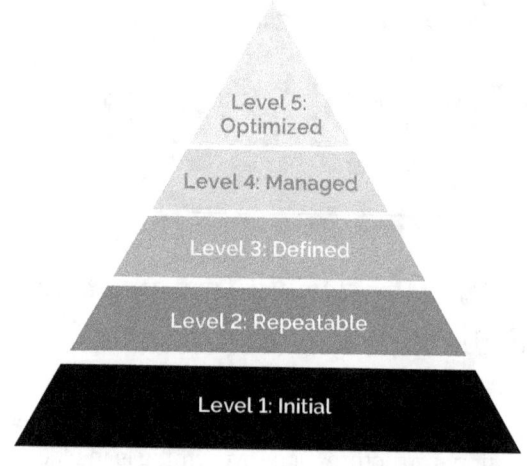

Fig.44. IT Governance maturity levels

IT Governance is a progressive process. In most organizations, implementation maturity is observable at the following levels:

Level 1 - Initial: Information stores are chaotic; Goals should be to discover and inventory content; clean up and reduce storage burden and gain stakeholder support.

Level 2- Repeatable: A small body of proven information practice and process is in place; Goals should be to reach a defensible position for key projects and departments.

Level 3 - Defined: Content is managed across project lifecycle; automated monitoring is set up for key policies; analysis, clean-up and metadata creation services are in place.

Level 4 - Managed: Information risk and quality support management reporting are in place; Value based content metrics are defined and measured; Information quality dashboards are in place.

Level 5 - Optimized: Practices for information reuse and value growth are in place; A master metadata connects and unifies content diverse sources to create new value.

Startups are at level 1 in the maturity model. The growth of the company will likely follow the maturity levels.

In risk management, you define policy and provide assurance in functions such as product development, finance, human resources, and quality assurance. Here, you have to implement policies and procedures to strategically reduce risk. These processes must include how and when you respond to breaches and issues when risk prevention methods aren't enough. Finally, you have to implement monitoring, including internal and external auditing to ensure compliance, and implement automated monitors or triggers to detect risks.

Proper IT governance allows you to focus on growth, because you know that risks won't be introduced as easily. For example, with password management in place, individual employees won't freely share passwords which will greatly reduce the risk of a data leak. Establishing the right processes upfront will reduce risks, while minimizing the total cost to introduce those processes later.

European companies also have to consider the GDPR, which stipulates additional IT governance requirements to mitigate and handle user data-privacy risks and issues.

In short, IT management and governance are a strategic business investment, which will pay off through long-term risk reduction and better data usage.

Access management

Access management issues are rampant in small organizations because you have limited resources to invest in individual user accounts and permissions. Instead, everyone uses the same source control, databases, and accounts. Passwords are shared with everyone to enable squads to rapidly complete tasks without waiting for access rights.

As you scale, this will generate risk because anyone can change anything. Most importantly, data in the platform doesn't likely belong to the company, it belongs to the customer. Offering everyone unmanaged access may be illegal under privacy regulations like the GDPR. Here, your most important risks include unauthorized access to sensitive data and data theft.

As Tech COO, it's your responsibility to mitigate access risks. Defining an access matrix is one way to do so. Here, you define resources available inside your organization, the roles in your organization, and how they link together. This access matrix then controls who can access, edit, and read various resources so that you stay in control.

Another important principle is to create user accounts for employees as a standard process. Here, it's important to use their name or an employee number, so you can easily refer to who accessed what and what they did. A password vault or tool is extremely important, because it will allow you to track individual user access.

This is important because many tools only support one user account per license, meaning you can't implement user management inside the tool. A good password management tool will store passwords and accounts centrally. This means it can track user access based on user-management, so that you can control access rights through password management. Password managers also allow you to track when and where each user requests access to software and files, which will help reduce access rights risks, so your organization remains compliant.

Set up a process to give proper access to new hires. Also ensure the process updates access rights as employees leave, so accounts are revoked to avoid unauthorized access to sensitive systems.

Change management

Quality IT Governance also means integrating strong product change management. This involves implementing risk management, deployment strategy, and documentation to manage the quality and impact of change. Here, you are managing risks such as teams making wrong changes that could break critical functionality, people with bad intentions (malignant updates), and loss of quality when updates don't maintain standards.

Change is an essential element of growing a SaaS product and creating structure and support to enable it is important. At the same time, change creates risk. A new update could crash your application or lead to data leakage. Customers might not like an expensive new feature. Or, features might not work correctly. You need strategy and processes to determine the direction of change, to encourage individuals to take responsibility for quality, and to monitor the quality of change. This is change management.

Change management often means establishing transparency in the development process and utilizing product owners and documentation, so that any change is traceable. Processes such as registering tickets and tying documentation and quality assurance into tooling will ensure that individual workers are aware of and adopt these practices.

You also want to integrate change management into the deployment pipeline, where practices such as automated testing, quality gating, and early validation are equally important. This ensures any changes going through to the live product have been validated and tested for quality.

Long-term change management is about tracking changes being made, by whom, and why, so you can maintain the quality of the product throughout change.

Incident management

Incident management is an important element of IT Governance because it allows you to ensure that incidents are closed, and to track and prevent them. An incident management policy should ensure that all incidents are registered and handled within SLAs. Incident management policies must include risk-management to ensure follow-up for incidents and to ensure root fixes.

You should also have monitors and triggers in place to warn of incidents, to recognize possible breaches, and to identify the level of escalation. Escalation, as discussed previously in this book, allows you to set priority for the incident while determining who should be notified and what should be communicated internally and externally.

Data security

Data security is an important and sensitive topic for SaaS products, which frequently store user data inside the product. Data security is important for complying with most regulations and law, but also to protect customers, prevent lost data, and protect your organization. This is important because most SaaS companies don't normally own data, they only store it. This adds responsibility in that your organization has to protect customer data. You need policies to ensure that data is handled with care to prevent risks.

Data security policies must define how you classify information (e.g. sensitive or confidential), how you protect that information, and how you store it. For example, how does your organization store client data? What about private client data such as credit card information? How do you respond in case of a breach?

Data is a very sensitive asset and it is a risk. Putting policies in place to protect data from loss, internal and external theft, misuse, or disaster is important to mitigate those risks. Some of these policies could include Information Security Policy or a Data Sensitivity policy to define internal and external standards for how data is stored, shared, and protected at different levels of sensitivity.

At Nmbrs, our IT Governance is implemented throughout a set of policies and work processes. Every organization faces risks which can compromise the quality and availability of the application. We maintain strict policies to prevent or mitigate these problems. Our primary risks include data loss, security breaches, the system going offline, or incorrectly calculated data.

Our IT Governance strategy focuses on minimizing these risks with policies visible to all employees. For example, when signing our labor contract, employees agree to comply to the following policies:

Access Matrix Policy – The Nmbrs Access Matrix defines resources such as servers and data and connects those resources to roles inside the organization. It maps roles to who should have access to certain tools, applications, and networks. It also manages access level, allowing us to create User or Admin accounts, giving employees access on a need-to-have basis. We periodically review access permissions during internal audit to ensure everything is up to date.

Data Security – Our Data Security Policy defines how data and sensitive information should be stored and accessed by Nmbrs employees. The policy includes, but is not limited to, financial information, employee information, customer data, documentation, application source code, and usernames/passwords or access data.

Information-Sensitivity – The Information Sensitivity Policy helps employees determine which information can be disclosed to non-employees. It also helps gauge the relative sensitivity of information and when it should not be disclosed outside Nmbrs without proper authorization. These guidelines cover information that is stored or shared by any means including electronic information, information on paper, and information shared orally or visually (e.g. via telephone and video conferencing).

Laptop Policy – This policy describes the use and purpose of working on a Nmbrs laptop, setting guidelines for security (e.g. hard-disk encryption, data must be backed up via cloud), and installation/maintenance of applications.

Not-BYOD – Our Not-BYOD policy ensures everyone receives secure company equipment such as laptops and mobile phones, to prevent the need from using own devices. It also includes policies in case employees access company resources such as email, calendars, contacts, documents, communication tools, performance tools, etc. using their own devices.

Product Development – Product development entails multiple risk factors, and we work to maintain ongoing strategy to minimize those risks.

- Product management includes several validation steps to ensure that only quality features are developed in the production environment
- The deployment process uses a pilot/co-pilot approach to minimize flaws/errors during the procedure
- Product QA is executed before every update to guarantee quality standards are met
- Employees use VPN connections and follow Data Security Policies to ensure the security of the systems they are working with
- Product changes are documented and traceable from documentation to code-level
- Different testing techniques such as unit testing and black-box testing are applied to all critical calculations processed in the engine

We also integrate other governance measures, such as an employee NDA. Here, every employee is asked to read and comply with our non-disclosure policy, as assurance they are aware of and ready to acknowledge their obligation to protect the confidentiality of the information they are working with.

These policies allow us to define and meet standards for quality and security because they are visible to employees, part of work processes, and integrated throughout the organization.

> **Strategy Tip:** Implement IT Governance to reduce security risks as you scale.

Maintaining Compliance

Most SaaS organizations start out small, typically with just a few people who are quite-often friends. Compliance isn't needed because you're not operating on a scale that would require it. As you scale, compliance becomes more important for several reasons.

Compliance is quite simply the process of ensuring your organization, its products, and its employees follow any laws, regulations, standards, and ethical processes applying to the industry. This covers internal actions and behavior (such as data security and privacy protection) as well as federal and state law.

Creating and enforcing compliance across your organization ensures the organization can legally operate, that it can uphold standards required to protect customers and employees, and that the organization is protected from risks such as data loss or theft, fines, lawsuits, and poor employee behavior. Compliance often starts with internal controls. These controls should be put into place to deliver a certain value or quality to the customer and therefore to maintain customer trust. These controls are validated by risk management processes such as compliance frameworks, internal audits, including the controls, monitors, and triggers in place to assess that risk. Having controls in place allows your organization to maintain standards of quality as you scale.

Here, compliance frameworks are one of the most important elements of quality control, because they create standards for your organization. You can choose to develop your own compliance framework or use an existing one. Many already exist, and many can be easily tailored to meet the needs of your organization. Control and compliance frameworks offer a lot of benefits in that they are aligned with the specific needs of different industries, take local and international regulations and law into account, and define standards for quality and performance. However, they will likely have to be tweaked or updated to meet your specific business needs.

COBIT and ITIL are among the most popular "out of the box" compliance frameworks.

COBIT – COBIT (Control Objectives for Information and Related Technologies) is a good-practice framework structured around governance and IT management. With a focus on maintaining quality of information, achieving strategy, and optimizing IT services in terms of cost and risk, while supporting compliance with laws, regulations, and policies, COBIT is very well suited to software companies. COBIT includes principles, best practices, tools, and models to guide internal compliance and auditing.

ITIL – ITIL (Information Technology Infrastructure Library) is a framework focused on aligning IT with business goals. ITIL divides business processes into 26 items, split across 5 stages, which each include actionable ideas for quality auditing and assessment. ITIL works to foster alignment between IT and business goals, to reduce costs, to improve the transparency of IT costs, assets, and risks, to improve business risk and service management, and to develop a stable IT environment to support the organization.

Having compliance frameworks in place will help guide your organization through external compliance as well. For example, audits are typically performed externally once per year. Developers often have no idea why or what they need. If a compliance auditor asks a question, developers cannot answer because they aren't involved in the process. Here, compliance frameworks can help to bridge this gap, so individuals and teams know what they are responsible for and why. Every compliance or risk-management control should connect to a team, a role, or a chapter, and they should maintain ownership of it. Then, when external auditors have questions, teams can answer those questions because they know what is going on, why, and why it's important.

Compliance and control frameworks are important because they offer the controls and change tracking mechanisms to quickly and easily set up internal controls and audit processes. These controls are also often used by external auditors, especially those checking for ISO/ISAE standards and so on.

Internal audits should be established as a subset of your full framework. These audits should be, in essence, a smaller-scale version of a full audit and should focus on specific important or high-risk items. Teams can integrate them into monthly rituals, simply performing the internal audit monthly or after completing an epic. Once put into practice, you'll be able to perform audits quickly on a small scale, so that you can manage the quality of risk management. When big audits do happen, you'll already know everything is up to date.

Here, internal audits serve a dual purpose. They must assess whether internal controls are adequate. They also have to determine if controls actually address necessary risks. Integrating auditing processes into standard business practice is important for maintaining the quality and validity of risk management inside your organization.

Here, it's important to follow Agile practices, aligning risk management processes into every level of organization. Operations and management must be able to perform organization-wide risk assessment and auditing. Teams must be able to do so on their own as well. Empowering everyone to assess their own risk will create a culture of informed decision-making, where everyone is aware that "risks" or "what might happen" are part of decision-making processes.

My recommendation is to ensure that internal auditing contribute to risk-management decisions by offering assurance, validation, advice, and insight, rather than simply rating something as good enough or not. Doing so will impact your organization as a whole.

Compliance efforts offer numerous benefits across your organization. For example, they work to maintain quality of work, quality of code, and how data and information is handled. Compliance helps you stay in control of how work is handled, which allows you to offer consistency to customers. This will eventually improve operational quality which naturally results in a better product and will improve customer trust.

Strategy Tip: Assign an owner to each control in the compliance framework to ensure continuity.

Compliance in Agile Environments

Agile is quickly becoming one of the most popular approaches to software development, but many organizations still believe it doesn't work in situations regarding regulation and compliance. With many industries facing strict compliance requirements, developing compliant development processes is crucial for achieving standards, meeting internal and external regulation, and sharing processes with internal and external stakeholders. Compliance is necessary for nearly any scaling organization because it guarantees your ability to evaluate or control the quality of your product or how changes are made to that product.

Agile makes compliance more difficult than traditional waterfall development because teams and squads have significantly more control over their own processes. They develop what's needed in a way that meets the situation, not in pre-agreed and compliant ways. This can make establishing Agile compliance more difficult. However, it's not impossible and you should take the time to implement processes to ensure compliance.

Auditors want to see that you can explain why and how changes are made. This might mean trace source code changes to project management tickets (or any other sort of ticket) so that there is a recorded reason for the change. This sort of linking is standard in software development and required for any sort of compliance or certification.

Compliance is also relatively easy to implement in Agile using processes. For example, squads can create a backlog and then work from that backlog as a project ticket rather than autonomously performing manual updates. This creates an easy-to-follow change log which you can use for compliance and project validation.

Testing and validation are other major issues for compliance. Most auditors will require proof that code validation and software changes are handled by someone other than the original developer. You can ensure this sort of compliance by automatically registering testers and developers in tooling, so that you don't have to process this sort of record manually. Here, you also want to automatically register testing and validation performed on code through the development pipeline, so data is available for compliance.

While automation is useful for compliance, you also need a clear manual approval process. Auditors want to see that changes have been approved. This often directly contrasts with Agile environments, where you need to empower teams to make changes themselves. In continuous delivery and DevOps, you have to empower the developer to make these changes for themselves. How can you balance that with compliance?

My recommendation is that you use processes to automatically register everything, so you can always prove who did what and why without making a very strict development process. Here, you can connect source code to tickets and requests and can show that certain changes were completed by the teams who own those products. This can also create a sort of auto-approval under company policy, where the product owner (the team) approves the changes, therefore meeting compliance requirements.

It's crucial to implement basic standards for continued analysis of your regulatory landscape. These standards can differ significantly from one industry to the next, with some industries such as finance requiring significant amounts of compliance. Agile and DevOps do not emphasize the upfront analysis and risk management required by these industries, so it is important to review, understand, and implement what is required for those standards in as non-constrictive a way as possible.

Some elements of Agile compliance are fundamental to DevOps. For example, establishing greater communication across teams with improved transparency and record-keeping ties into compliance. Similarly, leveraging DevOps and creating an automated pipeline ensures information is always available for internal and external stakeholders, so everyone stays informed.

Automation connects business, development, and operational units at early stages of development, automatically runs testing to improve product quality and reduce late-stage bugs and flaws, and helps create visibility and transparency throughout the pipeline.

This also ties into incident management, which is often strictly regulated by compliance and auditors. Here, you need internal and external SLAs so you can show who responded, when, and in what way and whether that did or did not meet your external agreements with customers and stakeholders.

Finally, your compliance requirements should be recorded, registered, and stored in a central database where individual developers can easily access them. Like other business processes, these requirements should be integrated into tooling, so they are followed as part of completing work, rather than adopted as a secondary measure or forgotten. A database should record compliance requirements for access, security, data confidentiality, data availability, authentication, logging, and audibility. Your tooling should enforce these compliance measures whenever possible.

Agile teams very often disagree with auditors. This is natural because compliance regulations often directly contrast with Agile methodologies. At the same time, you need compliance. A good balance happens when Agile teams record processes, work to establish transparency for decisions and testing, and have clear ownership of the products they are working on. However, teams often forget to properly register work or don't understand why it's important. Implementing awareness sessions so that teams understand why compliance is important will help to resolve this issue.

Here, Agile artifacts often closely mirror traditional audit artifacts, in that user stories are similar to user narratives, acceptance criteria are similar to application controls, and so on. Auditors can easily use these factors to evaluate compliance inside an Agile team without insisting on strict or rigid compliance processes.

Strategy Tip: Integrate compliance as part of Agile with automated reporting.

Continuity of Service

Continuity of service is the concept of having strategies and scenarios in place to ensure customers continue to receive access to software and tools if something critical happens. If something goes wrong, you have to be able to respond in a way that meets SLAs so customers can continue accessing the software they paid for.

"Critical" could imply a natural disaster physically destroying your servers, a hack, lost customer data, infrastructure outage, or any of several other critical incidents. No matter the incident, the result is that the customer can no longer access your software. It is your responsibility to fix this.

Creating a continuity of service strategy means conducting risk analysis to determine what can happen and how likely that is. You can then conduct an impact analysis to determine what the likely impact of each event is and how broadly it will affect customers and users in the case that it does happen.

Your continuity of service strategy must be scoped to meet the most pressing customer goals. These goals are often as simple as "meeting basic levels of service as agreed in the SLA". This means creating strategies such as a backup plan for the infrastructure or a separate server for the software which you can push live in case something goes wrong.

In other cases, your continuity of service strategy may have to tackle market factors. For example, if the market changes and your technology is no longer compatible with a key partner. You may want to define a process to quickly adapt your software to remain compatible. Or if your business goes bankrupt. You would have to create a mitigation strategy to ensure customers continue to see value. For example, using an escrow service to ensure software remains up and running.

Your primary question is often, "How can you make the application available to customers as quickly as possible after each incident". What steps would you have to take to ensure that the application would still be available? How can you reduce the risk of this critical event happening in the first place? How can you prevent the critical event from happening again (when applicable) once you restore functionality to your software?

In most cases, your strategy should include three basic elements; disaster recovery, documentation, and validation.

Disaster Recovery – You need an implementable solution to restore functionality in the case of each identified possible critical event.

Documentation - Automated documentation is crucial to identifying what went wrong, where, and how it happened. You should also document the processes and steps of any disaster recovery plan, so you know when, where, and how steps are completed.

Validation – Any recovery plan must be validated to ensure it works. You can achieve this through testing on separate servers or branches, so that when a critical event hits, you already know your backup measures work.

While you may be tempted to assume that low risk means problems won't happen, critical events hit even the largest companies. A critical failure will have low impact for your organization while your customer base remains small. As you grow, it impacts more and more users, creating larger and larger problems. Developing risk-mitigation and emergency reaction processes now will give you the tools to respond as quickly and efficiently as possible when something does go wrong.

Data is your organization's most critical asset. You need to ensure you can recover it in case something goes wrong. Defining a data recovery strategy is a crucial part of your data loss risk-mitigation because it ensures you're ready for worst case scenario. Your recovery plan should include a data backup plan, with long-enough retention periods to cover all risks. These backups should be stored safely, on a separate server than the primary database. It's also important to define a regular schedule to test backups, so you know you can restore them when needed.

Failing to create a good data recovery strategy is a mistake many entrepreneurs, including myself, have made. WallyLabs, one of my initiatives, had a relatively small customer base. As a result, we never really invested much time or effort into IT governance. Then, we were hacked. The hackers entered the database server and were able to encrypt the entire disk, where we had also saved out backups! From that day forward, I took backup strategy very seriously, even for small projects.

It's also important to keep in mind that organizations of all sizes suffer from outages relating to IT governance. You need a strategy to minimize and prevent risks whether your organization employees three or several thousand people.

On October 16, 2018, the homepage of YouTube displayed either an error message or a blank white screen. Clicking on user pages resulted in Google's classic "500 Internal Server error" message. The outage, which lasted just over an hour, impacted the site globally, resulting in a trending hashtag (#YouTubeDown) and hundreds of thousands of comments online.

While YouTube responded quickly and fixed the issue within an hour, the organization, which is owned by Google, is one of the largest in the world. Despite that, they still experienced a critical event affecting millions of users.

Your customers need to know that they can keep using your service, even if something goes majorly wrong. You have to define processes and strategies to solve these problems. You can then use this data to define what you will do to resolve problems and how long those fixes will take in your SLA. Most importantly, when something does go wrong, having Continuity of Service processes in place ensures your teams know how to react.

Strategy Tip: Create a continuity of service strategy linked to SLAs.

Conclusion

Proper governance and compliance will reduce risk, improve your ability to grow your organization, and enable you to earn compliance certifications across your industry. It's crucial to integrate governance and compliance as part of your culture, where individuals are aware of risks and mitigation factors, are able to contribute to solutions, and are part of regular internal audits.

Developing and maintaining clear policies on important topics such as backups, security standards, and recovery will help you to achieve this. Defining these standards ensures you have solid foundations and that teams understand what to drive for in terms of technical solutions, so they can focus on value delivery.

Alignment/Working with other departments

"Synergy - the bonus that is achieved when things work together harmoniously."
– Mark Twain

Departments such as tech, HR, and finance have historically been separated, largely because each required completely separate processes and people. Today, digital automation and customer journeys have brought many of them together, intertwining most aspects of traditional business. Previously, you'd hand a sales rep a brochure, your tech team would build the tool, and your finance team would send invoices. Today, everything has to work together. If you want to automate customer lead generation, conversion, a trial, and onboarding, you must develop sales into your product. If your customer journey isn't implemented around your product, you cannot scale.

Automation requires mapping the customer journey inside the product, creating touchpoints to convert and validate, and automating this process to connect marketing to sales to customer support and finance. Doing so means understanding your organization's customer journey and touchpoints across operations, so you can bring them together and move the customer seamlessly from one to the next.

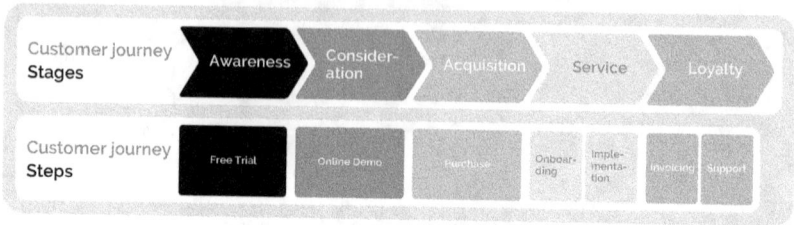

Fig.45. Example of Mapping a Customer Journey

For example, growth hacking is one of the first points of the customer journey. This involves generating leads, driving those leads to the website, and converting them to a free trial. The process is very technical but also necessitates connecting the website to the product, so customers can move seamlessly from one to the other.

Services must be built into your application platform, finance shouldn't be sending invoices, the product should do that. This will require a significant amount of interaction across platforms. In this chapter, I discuss the organization and structure necessary to ensuring teams can work together.

Sales and Marketing

Sales and marketing teams are normally outside of the scope of the Tech COO. At the same time, it's important you involve both in your organization's growth. Both are responsible for direct interaction with new customers, developing the customer base development teams are supporting, and are a good source of information for driving future organizational growth and application development.

While both are traditionally separate teams, with a divided customer journey and separate goals, this has changed. Marketing previously handled a small portion of the journey before handing it off to sales, who handed it off to finance, and so on, but today, there is now much more overlap in that both teams work with the same customer base. You have to bring sales and marketing processes together, into your SaaS, to make this work. As a result, it's very relevant for the Tech COO to be involved with both.

How does this work in practice? Sales and marketing massively impact the organization and consumer. Both connect with organizational goals and often make growth possible. At the same time, marketing and sales both rely on consumer behavior, data analysis, and customer feedback. This data is most often available through operations and directly in your application. Operations has the data to ascertain how and where customers receive value from products, which can then be fed back into creating more targeted sales and marketing campaigns.

Operational goals should also be used to formulate shared marketing plans, sales strategies, and campaigns. Marketing is responsible for acquiring leads to help company vision become reality. They must market in line with what the organization intends to achieve, what the organization is producing, and what the organization is able to deliver. If leads that marketing acquires need something your organization might produce or intends to produce, those leads won't be satisfied now. Gaining value from marketing means aligning teams and strategies with business objectives, with development, and therefore, with operations.

Customer information is likely scattered across multiple apps and therefore difficult to reach and organize. This logically creates an information gap, preventing customer service and development from using information from sales and marketing and vice versa. One strategy to ensure everyone remains on the same page is to simply keep all customer information in one place, preferably in the same application. For example, you can tie customer support information into every relevant team using a Customer Relationship Management (CRM) system or Customer Success Management (CSM) system. This isn't always possible for either team individually but can easily be achieved when both align with Operations to achieve a single goal – delivering value to the customer.

Doing so will work to ensure Operations is using customer onboarding data when setting goals and strategy, that support knows what information customers were given when they were onboarded, and that sales and marketing are aware of customer satisfaction levels based on why users started using the app. In both cases, aligning sales and marketing teams with operational vision helps both deliver value to the customer.

Strategy Tip: Connect sales and marketing processes to your product.

It is a good idea to ensure marketing and sales are involved with development teams. When you do so, sales and marketing understand what's in the pipeline and when it will likely be launched, so they can have campaigns and customers ready for new features. This means linking marketing teams to product development cycles as stakeholders. Here, sales and marketing teams can see what is in progress or being developed and act on it. Sales and marketing can also create and maintain a backlog of features they would like to bring into the product based on expressed customer interest and demand, creating a feedback loop where customer demand results in change and new features.

If you do enable this sort of backlog, it's important to create a process where new ideas are regularly discussed with Product Owners. This process should include room to discuss new ideas, make the product vision fit, review the technical feasibility of the idea or feature, and plan for development.

Sales and marketing teams often need access to development and technical knowledge and teams. This can directly relate to processes such as onboarding, integrating into planning future marketing efforts and making promises to new customers, or to providing complex services in some applications.

Implementing this sort of cross-collaboration is a challenge for the Tech COO because each of these teams typically works in very different ways. Many work in their own tools, have their own work mindsets, and their own views of the customer. Connecting these departments means creating collaborative goals and processes, so that sales and marketing naturally interact with each other, and both naturally interact with development teams. Using standard processes such as mapping the customer journey, defining individual customer touchpoints with each team, and how customers are segmented will help marketing and sales to work together. Implementing monthly (or weekly) interactions between sales, marketing, and development teams to discuss ideas, features, and feedback can also be helpful for ensuring continued touchpoints and feedback.

Strategy Tip: Create a feedback loop to development from sales and marketing and vice-versa.

Direct connecting development to sales and marketing will solve several issues relating to how sales and marketing promise features to customers. For example, marketing and sales often want to discuss upcoming features and services with prospective customers. But, if marketing or sales promise a feature that's still in development, the ability to finish development can make or break a sale. If the feature is never finished, you will have dissatisfied customers.

Teams often advertise based on that market, promise to sign contracts when features are delivered. This works in waterfall organizations but not in Agile.

In an Agile organization, it's almost never a good idea to promise undeveloped features, simply because you don't know what will be built. Instead, you have a backlog of features you intend to work on, which could change as the application, priorities, and even customer demand changes.

Connecting sales and marketing to the development process will drive more awareness of this process. Sales and marketing representatives can see the backlog and might have some idea of what is in the works but won't be able to share concrete plans with leads because there aren't any.

Sharing the backlog with customer service may also solve other problems. For example, customers sometimes pressure customer service for delivery dates for products or features. With a backlog, customer service will understand that they can't share a deadline because there isn't one.

Sales and marketing representatives will appreciate having a roadmap of intended releases. Agile work practices mean that roadmaps of any sort will quickly become outdated. Instead, sales representatives can create a general vision of the direction the application is going, in line with the backlog, rather than sharing a list of new features. Prospective customers can get an idea of what is being prioritized but should be informed that it can change.

Development can also choose to maintain a customer-facing backlog, so existing customers always know what's being worked on, what's been scrapped, and how development is going.

My company solved these and other problems by organizing commercial squads around the customer journey. We integrated a growth squad responsible for lead generation through the website. We also integrated an onboarding squad, responsible for developing tooling for user onboarding. Another squad managed events and conventions, where they would showcase our product and engage with people on relevant topics, to generate product awareness and interest. These squads combine commercial and technical people, allowing them to deliver a full solution to increase user conversion and satisfaction. These squads also had to connect and integrate CRM and chat tools, and the HR/Payroll product, so they could implement processes and workflows around the customer journey.

Our results speak for themselves. One of the most important was that these changes removed dependencies between sales and tech. Sales teams already had technical people who understood sale's needs, allowing them to develop their own internal products without dependencies. Their internal knowledge of what the sales teams needed resulted in a much-needed improvement in quality.

We also found that integrating sales with the product meant sales had more insight into customer data. Previously, they only had access to data collected through the sales pipeline. Accessing data through the application meant they had a better understanding of why and when real customers were using the product, allowing them to better market the application and target better customers, which improved conversion and customer satisfaction.

Finally, organizing sales around the customer journey clarified responsibilities, allowing us to seamlessly move customers from sales to onboarding.

Strategy Tip: Never promise features that are not in development.

Customer Support

Customer support is the one team inside your organization that consistently connects with and interacts with users. Most importantly, customer support interacts with users at your most vulnerable stage, when customers want or need something, typically because something has gone wrong. They receive customer feedback, hear user wants and needs, and are the primary customer touchpoint after sales and marketing. Customer service representatives understand your product, its bugs and flaws, and its difficulties better than anyone in your organization.

Their input, information, and customer knowledge are crucial to the rest of the organization. Yet, many companies fail to utilize that resource. Developing a structure where customer support is integrated into development, marketing, and every other aspect of the organization is crucial if you are to include customer feedback and experience into the future of your application.

Organizations are more-often creating this sort of cross-functional interaction, where product support members function as part of product development. As discussed earlier in this book, third line customer support is most ideal for this role, because most have the technical expertise to understand what development is doing while remaining connected to requests and feedback received through first and second line.

You may choose to integrate customer service into product development squads. More logically, you could create frequent touchpoints with product dev. Depending on the product, you may want to have weekly or bi-weekly touchpoints. During these touchpoints, Product Owners should discuss new features and planned implementations, customer service agents should bring up major product issues or bugs, and customer support representatives should cover major trends they've noticed in relation to feedback, customer wants and needs, and issues.

This will help product development to better understand their product quality and how it is received by the customer. It will also help product development to correct issues that are frequently brought to customer service, even if those issues are as simple as an overly complex function.

My recommendation is to implement weekly or bi-weekly meetings discussing problems and issues, monthly touchpoints sharing feedback and trends, and quarterly meetings to discuss the application as a whole. Customer service can share insights into customer priorities and wants, so they can add value to any road mapping or project planning as well.

Aligning customer service with development is important if development is to stay connected to what customers want. It's also important to create a channel from development to customer support, so that service reps are always aware of what's happening, when new features will be released, or the status for bugs or patches. Integrating both onto the same platform is the easiest way to achieve this.

> **Strategy Tip:** Align product development with customer service to create a feedback loop.

Finance

Few people think of finance when aligning an organization, but finance must be able to work with other departments. In a traditional organization, finance would invoice customers separately of nearly every other process. Today, invoicing is typically integrated with the application, customers are automatically invoiced, and the finance team must offer value in other ways.

Closely aligning finance to product development, sales, and operations will improve business results, as well as your product. For example, finance can show the value operations plays in achieving objectives including financial ones. The department can connect operational strategy to financial gain. They can also work to reduce overall spend or increase ROI for what is spent. This sort of collaboration is often hindered by top-down processes and a lack of mutual understanding, where teams literally come from different worlds and are unable to easily work together.

Here, finance and operations must align KPIs and metrics, create touchpoints, and align goals so that both work towards the same goals. Creating monthly meetings in which operations and finance align on goals is one important step. Another important step is to integrate operations into financial processes, so they have input on creating spend and budget and can better align with finance on ROI. Sharing at least some data across teams, so that operations can see spend and finance can see ROI is also a good idea.

Another example would be to align budget with technical teams. When Nmbrs migrated to a cloud platform with microservices architecture, we allowed product development squads to provision their infrastructure for components and apps. In our old environment, infrastructure was managed by a single squad and everyone was sharing Virtual Machines for development, test, and production. In the new model, teams were creating new resources without realizing costs, because they weren't accustomed to accounting for cost when making technical decisions. To prevent huge increases in budget, our finance team organized a monthly catch-up with technical squads to review their costs and to bring those costs into perspective with their work. We also introduced costs as one of the qualities to validate in Technical Designs, so that we could create more-cost-effective solutions.

Strategy Tip: Integrate costs into technical decisions.

Finance also integrates with development in that Agile budgeting is considerably different than waterfall budgeting. You could previously define a budget for a year and stick to a rigid total budget. Agile makes this difficult, because it's difficult to create a rigid budget or financial forecast if you don't yet know what you're developing. Agile uses point estimation to calculate what will be produced, so you can't precisely value what it will cost.

If finance makes budgeting decisions in-advance, teams can lose connection with a project and its costs. This naturally leads to problems in that you may pay a great deal for features offering little or even no value. If budget is spent on the wrong things, you'll have very little room to work with a sudden backlog or Just in Time (JIT) development. Agile development has to work with JIT, simply because it requires working on relevant projects as they arise. You have to be able to adjust budget in real-time, so teams can work on backlog items as they occur.

Connecting finance to development means that finance can always see what is needed and why. If you also connect customer support, finance can link budget to ROI through customer demand and value. This often means tying budgeting into the value stream rather than connecting it to a specific project. Here, finance develops budgets based on direct payoff or estimated value to the customer, rather than achieving a specific feature or functionality.

Finance cannot know what will happen in development 6, 12, or 18 months in advance. Instead, they have to closely connect with development and align metrics to understand where value is coming from. The Tech COO has to communicate and connect operations to finance, so finance has the overview to make decisions.

Earlier in this book, I discussed the idea that everything should be a product, which is important if you want finance to be able to track total budget across items. Here, you define modules, features, and other aspects as a product or as a value stream with a clear and definable value output, so that finance can budget for the product as a whole, funding it across squads and teams, rather than allocating a specific budget to each team or project.

Agile budgeting should be done in either monthly or quarterly sprints – with a running forecast. This will make calculating the expected projects and goals of the sprint and therefore the expected cost much easier. Because cost is usually directly correlated to the length of the project (set team costs) budget can often be the same for each sprint. Using this method, you can easily predict consistent sprint lengths and the number of work hours/team members in each sprint, which will enable your finance team to more effectively calculate the likely total cost of the product.

Here, you can take two primary approaches:

Burn/Cost Approach – Calculate total costs based on the cost per hour of team members involved in the project or the total cost of the team for the estimated duration of the project. Doing so means calculating the total cost of each employee, the total length of a sprint, and how that interacts with the project. For example, you will have to map how many sprints your team usually requires to complete an epic or develop a feature. This means having a good understanding of how long the project will take, how many hours each person will contribute, and account for additional costs such as hardware, third-party input, licenses, and unexpected needs separately. These can be accounted for based on need and an estimate or safety margin but must be accounted for.

Role	Annual Salary	Fully Loaded Cost	Loaded Cost / 2-week Iteration	Allocated Time on Project	Fixed Burn Rate per 2-week Iteration
Product owner	$70,000	$105,000	$4,000	100%	$4,000
Analyst	$50,000	$75,000	$2,900	100%	$2,900
Engineer	$60,000	$90,000	$3,500	50%	$1,750
Sr. Engineer	$90,000	$135,000	$5,200	100%	$5,200
Tester	$60,000	$90,000	$3,500	100%	$3,500
Sr. Tester	$70,000	$105,000	$4,000	50%	$2,000
					$19,350

Fig.46. Example with production and sprint costs

This approach directly contrasts with the standard Agile method of working with fictional estimates – such as story points – to discuss budget with Product Owners and Stakeholders. Instead, it allows you to quantify development costs based on actual work hours, which is much better from a financial perspective.

Precision-Alignment Approach – Precision-Alignment requires breaking projects or backlog items into specific tasks, estimating the time-involvement for each, prioritizing them, and budgeting accordingly. While it does require that you have a very strong understanding of development timelines, it allows you to break projects down based on must-haves rather than simply estimating for the whole project. This creates more room before you have budget for the full project but can take more time and require more involvement from developers.

In most cases, finance can use a combination of these two options to correctly determine expected budget. You should also take the time to perform similar budgeting with business case analysis, so you can budget in-house development and outsourcing, and compare the value from each.

Calculating the cost of a sprint will allow Finance to predict the likely cost of projects based on the number of estimated sprints. This then allows Finance to budget based on return on value, whether certain similar features are paying off, and by measuring the value received from customers through interaction with customer support. Keep in mind, there will always be additional costs. For example, hardware and licensing, unexpected reworking, unexpected problems, delays caused by third parties, or internal illness or disaster. Any budget should account for at least some of these factors so that when they do happen, it includes them.

Integrating Agile into Finance also means that finance will hold a retrospective with the squad or teams at the end of the sprint to assess whether the budgeting was effective and whether they needed more or less budget to improve the next sprint. This will mean assigning finance as a stakeholder on many aspects of development, so that they are present during kickoffs and major meetings.

Agile budgeting offers many advantages, including improving the efficiency and efficacy of budgeting, reducing risk in that mistakes will affect a few weeks at most, and likely reducing the total needed budget in that you only have to budget for a short period so minimal overhead and safety margins are needed.

It also makes room for choosing how and why finance operates. Here, you will have to work with finance to make a choice between zero-budgeting and incremental.

Incremental budgeting assigns a certain budget to teams at the start of a project or cycle and then uses what is spent as a starting point for the next. This has pros and cons. On the one hand, employees are held accountable when they exceed budget. On the other, they are often motivated to spend more so that they don't lose budget for the next project.

Zero-Budgeting begins each project or period with zero, effectively creating a new budget for each project. This requires more input from individuals and more communication with finance, which can lead to bottlenecks. However, it does force teams to justify costs based on sprints and specific needs, rather than a previous budget for a different project.

At a higher level, it's better to avoid bothering development teams with consistent direct costs issues. However, you do want to understand where costs are coming from and what returns you are seeing. This means understanding the cost per person, their time input per sprint, and what they are developing. Having this understanding allows you to move people around and allocate resources to squads providing the most value. When you have a fixed number of people in your teams, budget management becomes a game of resource allocation.

> **Strategy Tip:** Align Finance with operations and development to create a better overview of ROI.

Human Resources

Human Resources naturally intersects with operations in that both are about people and resource management. While the exact applications of each overlap a great deal, it is crucial that these two departments align and work together. HR works to supply the needs of Operations. When Operations expands a team or creates a new one, HR supplies the people to fill gaps. Both teams have to align, so they work towards the same goals, rather than struggling to manage who is actually doing what.

Operations and Human Resources have always been separate, despite being tied together in nearly all business environments. My advice is that you create a framework to align Operations and HR, with touchpoints that help both align on similar goals.

For example, both HR and operations work to ensure team performance. Both should continue to do so. However, HR should ensure performance through motivation and incentivization efforts. Operations does so through creating processes and structures that make performance natural or even possible.

In this way, HR efforts directly affect the ability of Operations to succeed. If squads are failing to meet production, you may be able to recognize that most problems are people-induced disruptions. If so, a solution would have to come from HR and not Operations. Integrating processes to improve productivity is crucial. These two must collaborate through shared processes, regular touchpoints or meetings, and close collaboration on goals and expectations.

> **Strategy Tip:** Implement core values and work to create a company culture working to achieve company goals and strategy.

HR and Operations are in direct conflict. Understanding these points of conflict and working to reduce them is crucial to ensuring both teams work together seamlessly. Operations Managers or Line Managers are often confused about what they are expected to do versus what Human Resources Managers are expected to do. For example, Scrum Masters in individual roles may struggle to realize which team they are in, but they are often in both HR and Operations.

One step towards solving this problem is collaborating with HR to clearly define roles and responsibilities. Here, Operations managers should understand their roles and where they intersect with HR and vice versa. Another important step is to create defined and transparent processes so that HR can see Operations processes and vice versa. This prevents conflicts of interest regarding responsibility and authority, so both teams can collaborate rather than compete.

A study by Cornell University, covering an anonymous Big Three auto company, reviewed a case where HR conflicts dramatically contributed to performance. The power-train facility had a history of poor performance, and, despite a strong emphasis on Lean, continued to underperform. In 2001, the plant introduced a new manager, who immediately recognized that many issues were people-induced disruptions. He introduced training to create mechanisms for recognizing people for successes, for recognizing bottlenecks, and for assigning these tasks to Lean managers or HR as necessary. The result was a massive turnaround on performance.[26]

Line and operations managers often manage employees based on total performance while HR cannot afford to do so. Ensuring that both teams understand each other's processes and priorities will avoid this sort of conflict.

Both teams need a significant number of touchpoints, including regular meetings, working inside the same processes and tooling, and the ability to have input on decisions made by the other team in order to work together.

Strategy Tip: Align HR and Operations and define clear roles and responsibilities for each.

Conclusion

Building an organizational structure can mean creating completely separate departments in silos, but this structure often does not work for tech organizations. Today's SaaS oriented tech companies must align departments around the customer journey to enable scaling and to deliver a seamless customer experience. This means ensuring that departments are aligned with the customer journey and organizational vision, that they collaborate, and that they are integrated into the application.

Departments must collaborate and work together, sharing data, becoming stakeholders in each other's projects, and closely aligning on goals, so they can deliver more and better value to the customer. Aligning teams with the customer journey allows you to connect those departments to the product, automate processes through the application, and improve the customer experience.

---

26 https://pdfs.semanticscholar.org/712a/753a0b2506d63555fb62818ab9e30a1b21cb.pdf

Continuous
Improvement

"An organization's ability to learn, and translate that learning into action rapidly, is the ultimate competitive advantage."

– Jack Welch

The environment in which your organization operates is complex and ever-changing. Market factors, consumer demand, competitor products, and base technology each influence what your customers expect, how you can deliver your product, and even what they need. If you want to continue to succeed in your marketplace, keeping up isn't enough, you have to innovate and offer continuous improvement and quality. Continuing to adapt, reinvent, and improve is essential to long-term success and must be part of your mindset.

It's a fact of life that nothing is ever final. This is especially true in SaaS environments, where operations and innovations must adapt to map a product decision or pivot a company. As Tech COO, you must be very open to this process of change and ready to recognize and act on opportunities.

Continue adapting

Technology changes at a rapid pace. In an average year, you could predict that the thickness of the average phone will shrink, connectors will change, and that we will once again double the number of transistors on a chip in accordance with Moore's law. The same holds true for software development, with innovative change appearing not as an exception, but as a standard. The influx of citizen developers, testing environments in the cloud, IT structures shifting to the cloud, and an increasing reliance on containers, AI, and algorithms as part of development are only a tiny fraction of recent changes to software development.

Change is immutable and unstoppable. As an organization, your goal should be to use that change, to navigate it, and leverage it to grow your company. If you do not, or do not keep up with change, you will be left behind. Competitors, even those new to the market, will quickly outstrip you if you don't continue to meet market and consumer needs. This requires consistent adaptation and innovation.

At the same time, adapting to change requires taking on risk. Innovation generates even more risk. You must balance the risk of change and innovation to prevent overstretching, too many costly failures, or misinterpreting the market.

Many organizations use some form of Kaizen, which relies on taking small and incremental steps to manage risk.

Kaizen is a standard practice in Agile, where organizations take on continuous but measured risks for managed improvement. Kaizen comes from the Toyota Production System, from the Japanese words Kai (Change) and Zen (Good), focused around waste reduction through continuous improvement and adaptation to change.

Fig.47. Continuous improvement cycle

Kaizen principles require taking small steps to reduce risk while consistently moving forward. The idea is to eliminate waste by reducing the potential risk, because a tiny step allows for very incremental risk and waste should you misstep.

This process of continuous innovation through incremental steps is a slow one. Any major change will require a significant time investment. Kaizen only allows you to build on existing processes. If you only follow Kaizen, competitors can and will outstrip you, often fairly rapidly.

If you're simply performing maintenance and incremental improvement, you aren't keeping up. You have to continue making innovative changes, or you will fall behind. If you only follow Kaizen, you can't take the big leaps that create "magic" – you won't be able to innovate.

Some organizations excel at using Kaizen and Agile or Lean to reduce waste and streamline development processes. At the same time, many need more than simple waste reduction to excel. Some of the most successful companies on the planet move beyond Kaizen to introduce factors relating to innovation, taking on massive risk that would never be supported by Kaizen alone.

For example, Pixar Animation Studios owes much of its success not to waste reduction and efficient production, but rather to its willingness to explore, take risks, and lose money when initiatives don't work. Here, Pixar Studios invests in 1-2 large productions every 3-5 years, investing in high-profile Kaizen techniques to reduce risk and ensure the quality of the end-product. At the same time, individuals are given the freedom to explore, try new things, and develop new techniques, all of which pays off in the form of Pixar's box office success.[27]

Ford Motors adopts a similar combination of Kaizen and innovation. First launched in 2006, "The Way Forward" combines Kaizen waste reduction, divestment, and improving products to reduce costs. But, simply manufacturing and improving existing models isn't enough to drive profitability. The Way Forward also includes a new program involving innovating products in-line with customer demand. The result has been an increase in production from 5.5 million vehicles per year in 2006, to 6.6 million in 2018[28].

What do these companies have in common with SaaS? They're both focused on consistently developing new products and bringing them to market.

While I believe that individual teams should use Kaizen to implement continuous improvement, there should be room for disruption and innovation.

As Tech COO, part of your role is to provide that room to manually disrupt and innovate. This means continuing to observe how teams work and then improving that in occasional big steps. It also means sharing and pushing for adoption, because without innovation, you won't be able to improve for the long-term.

What is innovation? McKinsey[29] identified 8 factors as crucial to innovation inside an organization, including:

- Regard innovation-led growth as critical and establish cascade-led targets to establish it
- Invest in strategic and risk-balanced initiatives with enough resources to succeed
- Develop market, technology, and finance insights to drive value propositions
- Create new business models offering defensible and scalable return on investment
- Launch innovations quickly and effectively to stay ahead of competition

27 https://rctom.hbs.org/submission/pixar-animation-studios-creative-kaizen/
28 http://cmuscm.blogspot.com/2013/02/adoption-of-lean-manufacturing-ford.html
29 https://www.mckinsey.com/business-functions/strategy-and-corporate-finance/our-insights/the-eight-essentials-of-innovation

- Launch innovations at a scale to meet your market and segment
- Create and capitalize on external networks
- Motivate, reward, and organize internal people to innovate

Each of these factors requires developing operational structure that makes room for innovation and large-scale change. This puts innovation firmly in the hands of the Tech COO, who must create that operational structure, designed around both risk management and innovation.

Innovation is the process of taking sudden and large steps, creating a new process. While both focus on improvement, innovation adds more risk, but also brings more potential for adding value. Innovation often has the goal of removing or replacing existing processes, rather than improving them. This naturally creates a great deal of risk, which, if left unmitigated, will cause decline.

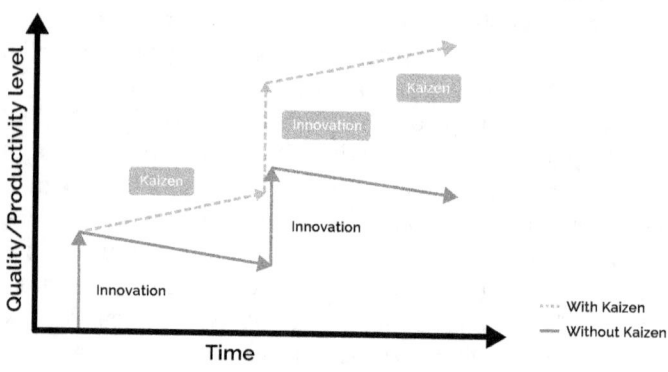

Fig.48. Improving with innovative steps

Your best option is to create a balance between innovation and continuous improvement, so that kaizen processes work to improve what you have until innovation creates something new.

At Nmbrs, we encountered similar issues through only utilizing Kaizen. Our teams were accustomed to working around specific country markets and to accountant and business customers. As a result, we had a Dutch business squad, Dutch accountant squad, Swedish squad, and so on. We optimized these teams using Kaizen to ensure they continued to perform.

At the same time, tech, sales, and marketing were very strongly linked. We decided to step in and make the innovative change of re-organizing them to align to the customer journey instead of markets, which the teams would never have completed on their own. Performing that innovation was not within the scope of the team. However, once implemented, Kaizen once again became effective in making it the team's responsibility to implement better changes and to optimize what that innovation brought them.

At another point, we wanted to migrate our application from a monolithic architecture to microservices. It wouldn't have been possible for every product squad to handle this disruptive migration on their own. Our squads were (rightly) focused on improving their modules. Instead, we set up a new tech squad to explore and try out the new architecture. Once the solution was drafted, the new squad helped the others to innovate on their modules inside the new architecture.

Both Kaizen and Innovation offer value, but one does not replace the other. You need both to continue to adapt and change so that you offer value to the customer while keeping up with your marketplace. Here, you can consider the "oil tanker" and "Speedboat" teams we discussed earlier in this book. Oil Tanker teams perform maintenance and continuous improvement. They move slowly, operate Kaizen, and focus on constantly improving what's there. Speedboat teams are all about innovation and disruption, they should be small, lightweight, and able to move quickly so that they can take big steps with a minimal amount of risk.

Strategy Tip: Balance Kaizen and Innovation to mitigate risks on all fronts.

Innovation Cycles

"The enterprise space doesn't move slowly because they're stupid or they hate technology. It's because they have users."

(Luke Kanies, founder and then CEO, Puppet Labs. Configuration Management Camp, Belgium, 2015)

Today's society makes new technology a fact of life. In a world where Moore's law has remained a constant, innovation cycles are a given. Software, technology, and even how we approach using technology is changing at a rapid pace and today's standard might be outdated in a few months.

Coping with a constant flood of new tools, frameworks, methodologies, and programming languages can be difficult. In some cases, products might be outdated before you finish implementing them.

At the same time, rapid innovation cycles are hugely advantageous for start-ups and scale-ups. Not only can you easily adapt to new technology and create new solutions and tools, you can move more quickly than large-scale competitors. However, it's important to keep in mind that changing technology can negatively impact your delivery, because any change in tools and frameworks will necessitate training or retraining and it will delay delivery.

Instead, it's important to create a structure where you allocate space and time for integrating new technology and adapting to changes, so that your internal processes and frameworks remain up to date without affecting delivery schedules. In most cases, this will eventually result in a Research and

Development team that can deliver new technology to the teams busy developing products.

Keeping up with innovation cycles means developing a process that ensures teams pay attention to and manage emerging technologies. You could assign someone to review new technologies, to research the state of technologies you are currently using, and to determine when new technologies begin to offer more than what you are using.

Conducting research, prototyping your options, and ensuring that anything you adopt is a good fit before adopting is crucial to keeping technology changes to a minimum. For example, you must take technical benefits, vendor lock-in, support, costs, maturity of product, community support, and other factors into account.

It's also important to pay close attention to product maturity before you switch to something new. Organizations have to migrate to new technology before frameworks, code, and technology are outdated and therefore negatively affect their product. It's also important to wait and allow new options to become stable before adopting them.

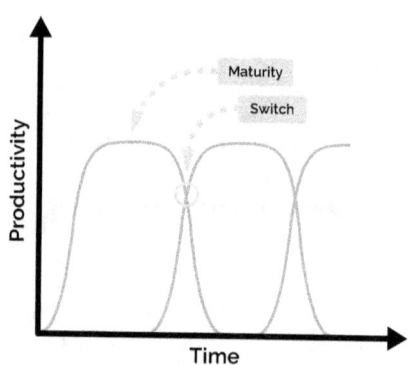

Fig.49. Technological innovation cycles

New technology is not a linear discussion. In some cases, technologies are greatly hyped, but they cannot provide the solutions or increases in productivity that you need. For example, some tech like Big Data, AI, and 3D Printing are immensely hyped, but decrease in popularity as organizations begin to understand the actual potential and possibilities of those technologies.

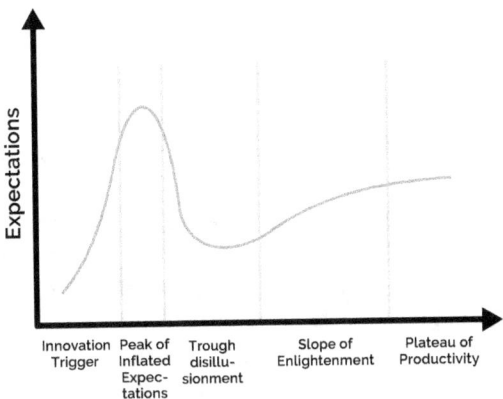

Fig.50. Technology trends hype over time

It's also important to realize when to leave technology behind. You don't want to continue investing in tech that will soon become obsolete. The later you switch to new technologies, the bigger the investment. This is because the technical gap is bigger and bridging it requires more development time to re-implement your product or parts of it.

To maintain your competitive advantage, you have to create processes to manage and stay up to date with technology. Doing so requires implementing a strategy to research, measure maturity, and measure the value of new solutions so you can move at the right time.

Strategy Tip: Dedicate resources to keeping technology and frameworks up to date.

The impact of change

Innovation and change are part of business development. They are necessary to ensure your continued success on a marketplace. Unchecked change and innovation can have negative business impact, in that people need time to adapt.

In 2011, Ron Johnson was hired on to J.C. Penney (American Department Store) in an attempt to copy his success at rebranding and revitalizing brands like Apple and Target. Johnson immediately set about changing store design and updating pricing, targeting younger customers, and taking on some $5 billion in debt. This shift alienated JC Penney's largest demographic – local, older, and value-driven consumers. JCP's sales fell by almost half and their stock plummeted.

While changes in retail management are a far cry from those in software development or presentation, the moral of the story remains the same. Even when change is needed, it creates a great deal of risk. Ron Johnson was later ousted from his position as CEO at JCP, but the company still needed major overhauls to complete the process Johnson had begun which was (needed) change30.

Organizations often implement change based on a foreseeable value or need, but then fail to make a full implementation of everything impacting that change, such as monitoring, processes, communication channels, team distribution, etc.

Here, Lego, the toy company, is another prime example. The toy giant launched an innovation program in 2000, creating a great deal in terms of new ideas and change to both internal work practices and external product. With no change management in place, Lego nearly bankrupted. By 2006, the company recovered financial stability, and having learned from mistakes, launched a new growth initiative, this time focused on both growth and disciplined or controlled growth. As CEO Jørgen Vig Knudstorp put it, the goal was to remain "Around the box" and therefore customer wants and needs, rather than way outside it. The result, a 30% increase in margin in 2011 and a 34% increase in 2013[31].

Change has a massive impact on people, both inside and outside the company. Internally, if you create change, you have to implement the tools and processes to ensure that it implements fully. You have to give space and time to adapt tooling, processes, and work methods to new innovations, so that those innovations are supported and able to succeed.

Your goal as tech COO should be to introduce change with risk mitigation strategies and implementation strategies already in place. This means fostering top-down adoption, so that leaders and management act as a good example. It also means recognizing and approaching the human size of change, which is often resistance. For example, if you're introducing a new work method, highly skilled individuals and those whose jobs are being dramatically changed will likely feel threatened. Introducing training and learning programs as part of the new change would help to mitigate that.

Individuals should understand how and why change is being made, should have the ability to see how it impacts them for the long term and what they can do about that, and have the tools to properly implement and make that change. Fostering real internal adoption of new processes, ideas, and tooling also means creating ownership, making change about the individuals it affects, and creating strong and clear communication throughout the process.

Change management also means understanding how change impacts other layers of the organization, at every scale. Even a simple change will often impact unexpected parts of the framework, simply because everything is intertwined. For example, when a change is announced, teams may stop maintaining an existing feature, because they don't see a point in continuing maintenance. This can negatively impact other areas of the product or even the customer experi-

---

30 https://guce.oath.com/collectConsent?sessionId=3_cc-session_31b47302-6708-4864-9068-359954e833bb&lang=&inline=false&jsVersion=null&experiment=null

31 https://www.lego.com/en-us/aboutus/news-room/2013/february/annual-result-2012

ence for the period. Conducting a risk analysis and change analysis before implementing change or innovation is an important part of developing sustainable change.

Nmbrs chose to change how our teams operated in order to better align marketing with operations and sales, which switched our focus from market-oriented to one which mapped the customer journey. We worked with marketing squads to determine what that change should be and to drive buy-in before implementing anything. Everyone was informed and ready for the change, but when it happened, it didn't work. We had underestimated the impact the change would have on our tooling and processes, which resulted in many people being unable to work using the new methods because there were too many uncertainties and loose ends. We had to go in and create new processes and make new tooling available to fully implement the change.

> **Strategy Tip:** Wait to introduce change until you can push a full implementation with support and structure for the change. Don't underestimate the impact of change.

Influencers of Change

Introducing change is difficult and can often seem like fighting an uphill battle. The more established your organization, the more difficult implementing any sort of change will be. Long-term organizational change must affect the entire organization in that it must be adopted at an individual level in order to produce benefits.

Here, it's always a good approach to invest in coaching the most influential teams in your organization so that knowledge and work methods spread through leadership. People naturally follow their leaders and will respect changes more when leadership adopts them first.

Unfortunately, top teams, such as management teams, can show the most resistance to a new methodology. These team members must be ready and willing to embrace change and must be able to make time to learn and adapt to new work methods. This can be challenging in an environment with very senior and/or busy executives, but is often crucial to ensuring adoption.

When such influential teams don't embrace changes, other teams will likely not embrace them either.

Introducing change means introducing monitoring tools to measure how and where change is being adopted, that you push change to teams individually, and that you motivate or inspire change. This requires overcoming key issues relating to individual uncertainty of new work methods and processes, building confidence in those processes, and driving ownership of change for every team affected. Leaders can have a huge impact on this, even in flat organizations, because people look up to and emulate leaders.

Strategy Tip: Introduce new methodologies to executive teams first and ensure adoption before pushing methodology down.

Changing the wheels of a running car

Product innovation is a necessary part of moving forward. At the same time, innovation often means dramatically changing features, frameworks, or technologies powering your application. Doing so requires major disruption.

Few SaaS companies can afford to completely stop their product for the purpose of integrating new features. Pausing service for 24 hours is damaging when you're running a very small operation. As you grow, the impact of taking your application offline will increase. You likely have to introduce disruptive innovation in the app while delivering value to your customer, which can feel a lot like changing the wheels on a car – while driving it.

You'll always have to ensure the operational train is running when you make product changes. It can slow down and it may even stop, but those disruptions must be very temporary.

This creates a challenge in that you must change the technology behind delivering a core product, change or update a process while it's running, or re-organize core teams while they are delivering value. These steps will eventually have to be taken or you will lose long-term productivity, but how do you actually perform them without causing delays? Most SaaS companies take one of two approaches.

The first of these options is to simply stop operations, make changes, and then continue. This approach can be extremely risky depending on the time involved, customer load, and when you can implement change.

At Nmbrs, we mitigate this approach by conducting a risk analysis to determine the actual impact of the change. This impact is almost always larger than you'd see at a surface level. For example, if you're updating one aspect of the application, which only affects 1 team, you might think it affects one team. It's much more likely that if that team has to actually stop, dependencies mean most or even all your teams will have to slow down or even stop to account for those dependencies. This will have a huge impact on your delivery.

Therefore, I firmly believe stopping to integrate change is not scalable for a larger company. As a small organization, you can stop for the weekend, host a hackathon or similar, and implement change very quickly, but as you scale, this will no longer be possible.

The second solution is to simply branch development, using the same split integration discussed under Delivery Pipeline Automation. If you can roll new tools, processes, and frameworks to a duplicate of the running process, you can test its viability and performance before switching to the updated version. Here, you can create a new task force around the innovation track, so the team has space to try new things. They can work without the limitations of Kaizen, work without maintenance or extra load, and then merge back. They obviously have to stay in sync with other teams to ensure anything they update will merge back, but this can be managed by frequently merging back and forth, using toggle switches to hide unfinished updates, and continuing the process.

While you logically cannot duplicate teams, you can allocate teams or parts of teams to implement new technology while the rest keep the value stream running. Once each section has finished implementation on their end, you can allocate a new team, or begin roll-out depending on what you are implementing.

Once innovation teams have features or software proven to work, with proof of concept, you should merge back.

Here, you have two options. A "big bang" where you integrate everything at once with a short period of downtime, or introducing smaller batches of updates, so that big features move into the running train without stopping it. In the case of a "big bang", you will likely have to take a weekend to update the application to avoid downtime during busy hours. My recommendation is to use the series of smaller updates, which allow you to push even large changes with minimal impact to the application.

In April of 2018, Google inserted massive changes into their G-Suite, including tasks and a calendar integrated into Gmail. The roll-out, which included huge updates to security protocols and interface, went largely unnoticed. Millions of paying and free customers received their application over a longer period, as Google slowly pushed updates out with almost no impact for users.

This type of massive update, with little impact to users, should be something of an end-goal for most SaaS organizations.

Strategy Tip: Roll out innovations in small batches, such as per module or team.

Look to the past to make a better future – Retrospectives

Retrospectives are a crucial element of Agile, allowing teams, squads, and individuals to review past performance and actions for future improvements. These review processes allow peers to offer input on processes, quality, data metrics, and team spirit to provide input on what went well and what didn't. These sessions can then be used to optimize future sprints, improve actions and decision-making, and to optimize processes for better performance.

Unlike waterfall management practices where retrospective meetings involve management, Agile retrospectives logically take in feedback from everyone involved, including external stakeholders.

Retrospectives are held at the end of a sprint, event, or other interval, which should be clearly defined as part of your processes. Good times to hold retrospectives include after sprints, after epics, and after solving problems or escalations of any kind. My advice is that you also align retrospectives with important cycles or moments inside your organization.

This may mean connecting to a quarterly or yearly season. For example, you can retrospect the vision once per year, the business plan once per year, and the strategy once per year. Other, less fixed, elements of the business model may benefit from more frequent retrospection. For example, you may want to retrospect organizational goals as you complete each one.

You may define different retrospective periods based on your own organizational teams. It may also be valuable to allow teams to set their own retrospectives, so long as they follow general frequency guidelines. Retrospectives should be a ritual for the team, and that sometimes means they must be unique to that team.

No matter what you choose to do, retrospectives should align with important moments or goals, so that the topic and goal of the retrospective is clear to everyone involved.

During the retrospective, team members and stakeholders are asked to come together to discuss the previous period or project. Discussion should be held without blame, considering what went well, what did not, and what could be done better. The goal is, most often to improve, which requires analyzing work, tooling, decision-making, prioritization, and approaches. Teams can then decide as a group to make specific changes, giving them ownership of the process. It's important to set up processes that ensure everyone brings feedback and is involved, and everyone has ownership in the process. You also have to ensure that retrospectives are designed so teams have to create an action plan and follow up on that plan as part of the process.

Retrospectives have to happen at every level of the organization, at different levels of operation, and throughout the value stream. They should consider aspects such as strategies, roles and responsibilities, processes, metrics, and team satisfaction.

In my opinion, retrospectives are one of the strongest core values of Agile. They allow you to review the root cause of problems, what went right or wrong regarding the event, and how to make the process better for next time. They also allow you to continue to assess the value and direction of assets or goals, so that you can continue to improve and keep the organization on track, even as fundamental elements of market, operations, or strategy change around those goals.

My organization implements retrospectives as a ritual, performed at key moments such as when reaching goals. Here, we discuss success, what we could have done differently, and how we could have improved on our success. These retrospectives serve as learning moments, allowing us to review, improve, and learn from both our mistakes and our successes.

Strategy Tip: Set up a cycle of retrospection through every level of the organization.

Conclusion

Continuous improvement is no longer a "nice to have"; it's a survival strategy that will impact your ability to survive in a rapidly changing market. Standing still is today's equivalent of going backwards, because competitors everywhere are disrupting their markets, innovating, and offering new and better ways to do business. Succeeding in a world of change necessitates adopting that change at every level of your organization, from company culture to processes, so that you can innovate, maintain, and continue to improve.

You don't do it alone - Partners and Suppliers

"Alone we can do so little, together we can do so much"

– Helen Keller

While most organizations don't launch with the intent to partner with other companies, most eventually must. The nature of modern commerce means everything is connected and customers expect everything to be inter-connected. This often means developing integrations and partnerships as part of your operational model.

While many organizations start off with the idea of doing everything themselves, this is often neither efficient nor cost-effective. For example, partnering with suppliers who offer infrastructure and hosting (such as Amazon and Rackspace) is more sustainable and scalable than hosting yourself. Creating organizational structure to incorporate these partnerships is important if you want to create long-term, scalable, and sustainable relationships, benefiting both organizations.

Know your core, outsource the rest

Most organizations have core products and services. They are likely defined in the "About" section of the website, should be defined in company vision, and are defined in company goals. Core products are most of what an organization does, its goals, and likely the primary element of production.

Most organizations also eventually need or want to outsource some work or services. This can come into play during large moments of growth when you cannot keep up with demand, when new services or features are needed, or when your organization otherwise cannot provide for itself. Here, many organizations struggle with choosing what to outsource, when to outsource, and whether outsourcing is a good idea. While outsourcing can have a lot of benefits, my recommendation is that you never outsource your core products or services.

Why? Your core is what differentiates you from competitors. Your organization excels at your core processes, it's what you do, and it's what defines you. You invest a great deal in core products and services, and you understand how and why they work. When you invest in your core, you continue improving your organization and its output as a whole.

This doesn't hold true with products and services that are not core. If they aren't crucial to what you do, you will never have the resources or inherent knowledge to truly excel at them. This means you will have to invest a great deal to make even incremental improvements on these services. For example, if you're implementing IT Services to fix computers in your offices, you have to hire on a team, offer training, provide tooling, and invest to ensure they keep improving. However, because IT services isn't a core focus, work will be light and largely uninteresting. Your team won't have the resources or motivation to continue to improve.

If you were to outsource this service to a professional IT repair company, the result would be quite different. The team would have a range of projects to work on, a large-scale investment on training and development, and the ability to quickly meet new needs or wants.

Any work with a limited scope inside your organization is suitable for outsourcing, when it serves a support function. Here, commonly outsourced roles include print management, IT services, graphic design, accounting and payrolling (until your organization reaches a size where it makes sense to bring on a full-time accountant or bookkeeper), compliance, and many others.

There are several benefits to outsourcing. For example, outsourcing frees up staff to focus on core competencies such as software development or network maintenance. Bringing in specialists for one-off or very small-scale projects that aren't part of your normal scope of operations will speed up execution and likely improve quality. Finally, you won't have to finance any training because most partners will have access to information and training your team will never have unless you switch business focus.

Outsourcing can be a valuable resource but should be managed correctly. You must choose partners who fill the skills and training gaps in your organization to offer real value, rather than simply performing an auxiliary function you could do yourself. It's also crucial to avoid relying too heavily on an external partner. Make sure that at least some members of your own teams are trained in the work or process and can take over, even temporarily, should something go wrong.

Outsourcing isn't always the right decision. You may also choose to insource, where you disperse work among your team. For example, if your IT staff have some expertise in repairing computers, you may be able to add computer repair as a function of their role rather than outsourcing it. Insourcing is significantly less reliable in that you will still have to ensure employees have the training, tooling, and continued support to perform their jobs well. However, it will give you more control over how work is completed.

My company develops an HR/Payroll platform, so I consider our core to be product development for the platform, plus core functions such as the payroll engine and HR features. These make up our market differentiator. We also offer other modules, such as an integration marketplace. We needed a very good marketplace to fulfill our USP "easy to connect", but it wasn't something that set us apart. This marketplace isn't a core product, so we decided to work with a partner who could help us develop and integrate such a platform into our product. We also work with cloud hosting providers, because while it's very much important to us providing a stable product, cloud hosting is outside the scope of our expertise.

Strategy Tip: Outsource what is not core when you want to scale.

Choose the right partners

Choosing a business partner is a lot like starting a relationship. You have to get along, you have to have the same general goals and philosophies, and you need to be on an equal-enough setting that you can contribute to each other. This means paying attention to how your prospective partner works, what their vision and strategy are, what their services are, and what you can expect. It also means developing open communication channels and aligning your strategy.

This is especially important in terms of IT, where misalignment can happen quickly. For example, if your partner's vision creates a shift in focus, they may stop investing in the services you need. While this won't make those services immediately unavailable, it will likely affect the quality of service, your partner's ability to support new technology, and the future scope of work.

Here, the most important considerations are business mindset, business size, and business goals.

Mindset comes into play because it affects how work is done and how goals are addressed. If your organization is very Agile and modern, you won't work well with an organization that relies on top-down management to provide solutions for engineers to complete.

You also want business size to align to some extent. While it's okay to work with larger partners and to work with smaller ones, you do want a balance. If you work with a large supplier, you won't have the influence to create change your customers want or need. If you work with very small suppliers, they likely won't be able to scale with you to meet growing demand, unless they have other business drivers pushing growth at the same time. Size also impacts your ability to align quality, because a large company naturally has more time, people, and resources to invest in maintaining quality.

Your partner's business goals are also important, because you want to manage where they are going and whether that lines up with your strategy and plans. If your organizations are moving away from each other, chances are that eventually, the partnership won't work.

It's also important to pay attention to factors such as security. What are your partner's vulnerabilities and risks? How are they mitigating them? What about consistency in service? What are their SLAs? How do they maintain them? How do they prevent their errors from impacting your service? Backups. How do they prevent lost data? Scalability. Can they grow with you? Can they scale up if you suddenly increase customers? What about maintainability? Are they producing semantic and easy to maintain code if relevant?

Any partner you work with should have processes in place to maintain quality standards to a level that meets your needs. They should also be able to provide KPIs and other data proving they meet those standards. You will, of course, have to provide that data as well. You should also prepare SLAs, requirements, create procedures and processes for meetings, determine SCRUM prints, set contact persons, and define processes for Request for Changes upfront to ensure you have standardized ways to maintain quality with your partner.

Choosing your partners often means defining strict standards of what does and does not work with your organization. Once you do so, you'll more easily be able to determine if your partner can work with you, scale with you, and continue to support your business needs for the long-term.

When Nmbrs launched, we chose a hosting provider as a partner. Over the years, they delivered exactly what we needed and offered great support. We grew with them for a long time, offering input so they knew how to improve their service and which features they should add.

Unfortunately, we eventually outgrew them. Our demand for scalability or flexibility in our cloud infrastructure increased dramatically because of rapid growth. Our business model means that users peak during the last week of the month and first month of the year. Our weekend and out-of-office hours demand was considerably lower than standard as well. Our partner only offered monthly contracts and was not able to adapt our infrastructure capacity to our user needs. Instead, to keep up during peak use, we had to over-capacitate, incurring huge costs. Our partner was unable to adapt to meet our business model, because we were the only client with those needs.

We chose to diverge, selecting another partner who could meet our new needs. At the same time, there were no hard feelings, as our 10-year cooperation was a very important part of our journey to growth.

Business operations are another important consideration. How do you operate as a business and how your prospective partners align with that operation? Do you guarantee a certain level of uptime in your platform? If so, can your partner's services meet that SLA? Are your partner's risks aligned with everything you promise your customers? While your partner is not part of your service, their product availability likely impacts the functionality of yours in some way, or you wouldn't be partnering with them. Can you align with your partner on strategy so that they also maintain your SLAs and take responsibility for failing to uphold them?

If you have a breach of contract with a customer and they file a claim, can you forward it to your partner if they were at fault?

It's also relevant to consider if your partner can support your peak use periods. While important for most types of software, it's especially so for finance and other SaaS with peak use and low seasons. Can your partner effectively scale up and down to meet your volume needs in peak periods? You have to consider service and use fluctuations in your business model when choosing a partner or be able to align their provisions, costs, and scalability with your fluctuations. Can your partner meet peak volume use? Will your costs rise exponentially during peak use? Are you paying for peak use throughout the whole year? If you can align with your partner, you'll pay for what you need, manage costs in peak periods, and have a scalable model that is sustainable for both parties.

Here, the total cost of ownership is an important factor. If your partnership is providing a software solution or a service, you have to factor in the cost and time of maintenance, how much of that you're doing yourself, and how that impacts your customers. You have to consider how long it takes to implement new solutions or updates when you change the parameters inside your own software, because a partner who is unable to keep up will quickly increase costs.

Your partner also has to align on key areas such as growth. For example, if you're running a SaaS platform and outsourcing your IT infrastructure to a third party, you're reliant on them for growth. As your customer base expands, you'll have to invest in more IT infrastructure, more web servers, databases, etc.

Without proper alignment, your costs could rapidly outstrip your revenue, and you won't have a sustainable growth model.

Strategy Tip: Align your business goals and strategy with that of your partners so that they provide a similar scale and quality of product.

Their quality is your quality

A chain is only as strong as its weakest link. It doesn't matter how much quality you are providing if your partner isn't meeting that same quality. It's also important to consider that this metric goes both ways. If your quality falls behind, the total product quality is still only as good as the weakest link.

Your partner is likely intended to either fill a need or solve problems, but you are working with them. You have to invest to keep up with their quality. They have to invest to maintain or keep up with your quality. Here, it's important to put standards and practices in place to monitor the quality of service and product delivery. Everyone must be aware of what is being produced, how quality is measured, and how quality is maintained.

It's always a good idea to install your own auditing practices, so that you can validate both your own and external data. If you're receiving KPIs or other metrics as part of your contract, you have to be able to audit data and verify them. Here, you also want to review the quality of data, such as how often it is maintained, how often documentation is updated, which records are kept, and how it is authenticated.

Your partner should be able to commit to your SLAs, so you can commit to maintaining quality, providing new innovations, and creating real solutions together, rather than simply forming a temporary partnership and implementation.

No matter what your quality standards, you should have processes in place that allow you to share quality concerns and questions with your partners. It's also important to create retrospectives or similar sessions with partners, where you can periodically review quality with partners, either in connection with an important event or moment or on a periodic (quarterly or yearly) basis.

> **Strategy Tip:** Create processes to ensure your and your partner's quality remain on the same track.

One of the easiest ways to manage partners and their quality output is to create open channels of communication by dedicating internal people or teams to them. This can seem counterproductive in that most organizations don't outsource with the intent to keep maintaining or working on something. You likely want to outsource and have another organization handle 100% of that product or service on your behalf.

While this is understandable, you should maintain a certain level of control, so you know what is happening and why. Establishing an internal liaison with your partner is the easiest way to do so, in that you establish a role where at least part of that person's responsibilities are managing quality, understanding what partners are doing, and working to inform that party of what you are doing and what you need.

For example, if you are outsourcing IT services, they don't have the resources to proactively ensure everything is working in your organization. You need someone internal to recognize when there are issues and push your supplier to offer service. You have to create a communication and coordination point to ensure they cover everything.

Imagine a case where you want to outsource your internal IT services, to grant/revoke access to systems, deliver new computers, etc. You have to define a clear list of apps for the external party to manage access, in accordance with the SLA. This is complex because internal IT boundaries can be blurry in that they interconnect with procurement and finance, or device policies and HR. Moving responsibility and coordinating these groups to a third party can be very complex. Having an internal liaison or team to handle services on your end will resolve this complexity.

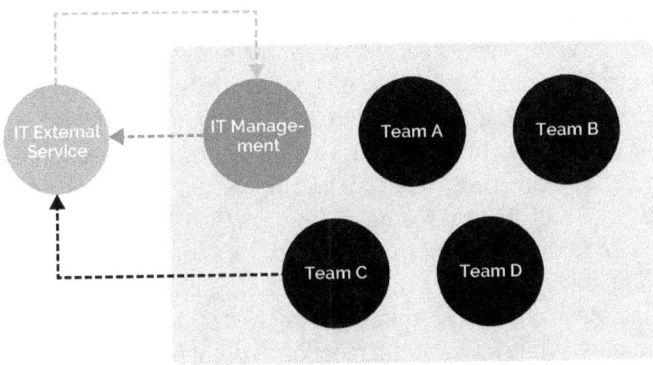

Fig.51. Internal liaison when working with partners

This also holds true with internal communication. It's often important to have an internal representative, so that developers and teams know who to contact when they either need something or feel as though the outsource company is not delivering. While it's a good idea to set up direct contact through phone, ticket system, and email for relevant teams or services, having an internal liaison makes the process easier. This liaison can then be responsible for service on an internal level, so they are responsible for measuring quality and value, as well as holding the partner to their commitments. In an ideal situation, any outsourced service would automatically be problem free. However, partnerships are an investment, and you should measure the quality and value of any partnership.

Team requests will sometimes fall outside of the scope of services provided by a third-party. Here, an internal liaison or team will establish the option for the service-provider to communicate these requests to an internal and responsible party instead of denying a service. Here, the liaison can either choose to handle the request themselves or extend the services of the external party, allowing for more options inside the organization.

Managing third-party relationships through an internal liaison will allow you to create a single point of communication, quality management, and service approval, which will improve communication and quality of service.

Strategy Tip: Keep an internal liaison for external parties.

Clear boundaries

The more you work with partners, the more likely it is you will experience either overlaps or friction in service. Here, it's important to set clear boundaries, so you know who is responsible for what, so that every element has a defined product owner, and you know where to turn when something goes wrong. These boundaries should be part of your contract.

If you are partnering with an organization to offer an integration to your users, you likely experience a great deal of overlap. The teams who own the implementation in your organization will likely butt heads with the teams owning the implementation in your partner organization. It's important to define who should be doing what, so that each organization is aware of their own role in the process.

What are they delivering? What are you delivering? How do you ensure you can deliver your half of the agreement? How are they ensuring they can deliver on their part? Taking the time to create very clear deliverables and SLAs will pay off in the long run.

One of the downsides of outsourcing services it that it often creates an "us vs. them" mentality. While it's a natural reaction to point fingers at an outside party when something goes wrong, this reaction can be harmful. Defining clear responsibilities ensures that when something does go wrong, you know who to contact and who should fix it. This will avoid unnecessary "finger pointing" where you look for someone to blame, and really just push blame at the third party.

You should also have strict procedures in place to ensure that liaisons and teams working together between organizations understand what their boundaries and responsibilities are, so those team members can work together without friction. At the end of the day, your goal is to create quality service for customers. Together, you want to produce something better than you can alone. You need to be on the same side with clearly defined boundaries and responsibilities to do so.

> **Strategy Tip:** define clear deliverables and service levels with external parties.

Conclusion

Working with third parties and partners is almost a given in the modern software environment. Whether you choose to create third-party implementations through apps and plugins, directly integrate into another software platform, or are outsourcing support services such as IT, partners will come into play.

Establishing good relationships, aligning your strategy and goals, and working to choose partners who can meet your quality and scaling needs will help you to get the most from these partnerships. It's also important to set boundaries, share SLAs, and agree to common communication standards through specific points such as liaisons, regular retrospectives, and frequent communication.

FINAL NOTES

Over 1.5 million tech startups launch each year but most fail within the first 5. It's well known that startups fail at a massive rate, but why? My opinion has always been that failures often relate to gaps in operational structure, which impede an organization's ability to grow. Unfortunately, many organizations simply don't have the experience to understand what they're missing or why it's needed.

Seed programs and accelerators give these startups the necessary funding and information to create a minimum viable product, but few tackle the next steps.

Where do we go from here?

If even excellent products fail, what can we do to scale successfully?

How do we support growth as product demand increases?

Studies by companies like StartupGenome suggest that 70% of startups fail because they scale prematurely, without the structure or strategy to enable growth.[32] This aligns with case studies, like that of Pets.com, which raised an estimated $290 million with seed funds, Amazon investment, and a move to public offering. The company still failed within 3 years of launch, thanks to factors such as no market research, poor strategy, and over-investment on driving customers with massive advertisement campaigns, rather than focusing on first building the foundations for growth[33].

Bridging the gap between startup and scaling is where many organizations fail. As an entrepreneur and the co-founder of a startup myself, this gap hits particularly close to home, but it is one I have successfully bridged.

When I co-founded Nmbrs, I had no idea where my journey would take me. Today, I've successfully developed an organization that now employees 120+ people, serving software solutions to thousands.

I developed the Vision to Value Framework at Nmbrs over the course of several years of experimentation, and change. It is a model that emerged from the experience of implementing different frameworks and processes, developing a clear vision and strategy, defining work and scope, and developing processes for teams to deliver and work.

---

32 https://startupgenome.com/report2018/
33 https://brainmates.com.au/brainrants/pets-com-a-classic-example-of-product-development-failure/

1. Strategy (Why)
2. Work management (What)
3. Process management (How)
4. People and Structure (Who)
5. Data and information management (Bringing the pillars together)

These 5 pillars will help you to build an organization capable of producing the 7 delivery qualities. These qualities, which include product quality, service quality, development velocity, team motivation, innovation, cost-effectiveness, and compliance are the eventual result of designing your operations in a way that delivers the qualities your organization needs. Because, at the end of the day, operations is about producing quality results.

Each pillar of the framework serves to support the other. This means you need a clear vision and strategy to support growth, but without defined work processes and organized teams to deliver on that strategy, it has very limited value.

This model is often applied invisibly, with no specific tools or processes in place. It was my responsibility to monitor the organization around the 5 pillars, identify bottlenecks, and take action to improve. Once I achieved balance, I would see increases in productivity, engagement, and motivation from both teams and management. When any one pillar is not optimized, I could see a lack of direction, chaos, confusion, and lack of clarity. Balancing the pillars is what allowed my organization to scale and grow in a structured way, which continues to support us today.

I am sharing my model with the intent to guide future organizations on their journey. Use it as a compass to drive your organization, so that you know what and where to look for problems. Consider it a tool in your tech COO toolkit, which you can apply in many environments, regardless of the type of product your organization develops.

My model applies to startups as well as larger, scaling organizations, but your implementation will be part of your own work and journey as operational lead. No matter how you apply it, the principles will remain the same. As you apply it, I would strongly recommend implementing control processes so that you can optimize and improve it over time. For example, consider doing an organizational X-Ray every 6 months to identify new bottlenecks, verify that previous ones were handled, or to review that items are being handled in the most optimal way.

Today is an exciting time for organizations of any size. My operational framework is simply a tiny piece of a much bigger iceberg. Organization design is growing as a trend, people are expanding into new and exciting leadership methods like Teal and Holacracy or Flat. The simple process of how we do business is changing. For new organizations, these changes translate to opportunity. You have the opportunity to explore, to reinvent and to create an organization that truly reflects your organization's goals and ideals in ways that no one has ever done before.

It is my hope that this book will help you on your way to greatness, helping you pave the path that leads you to growth. This is your journey from vision to value, go make your mark on the world.

Acknowledgments

I would like to extend my thanks to the many people who offered input and participated in the creation and writing of this book. I could not have done it without you and I am grateful for the opportunity to learn and grow with your help. Writing a book is not an endeavour to do alone. The people who motivated me to keep going, who spent precious time reading drafts and providing feedback, and who gave me advice and guidance on the content and the publishing process truly made this possible. They are, in random order, Brandy Burden, Michiel Chevalier, Mariel Drommering, Justine Broekhuizen, Bianca Prodescu, Jan van Beersum, Floris Drost, Redouan Aouragh, Julia Hack, Nathalie Odermann, Rita Gomes de Abreu, Joao Gomes de Abreu, Charlotte Scott-Wilson, Egbert Clement, Amanda Cardinale, Charlotte van den Brekel, and many more whom I would not have the space to list here.

Great thanks to Nmbrs, and my co-founder Michiel Chevalier, with whom I worked for so many years to gain the experience I share in this book.

Special thanks to my partner, Ana Maia Marques, who gave me the energy to start the project and continued motivating me to finish it.

And finally, thank you for reading my book.

www.ingramcontent.com/pod-product-compliance
Lightning Source LLC
Chambersburg PA
CBHW070526220526
45467CB00003B/876